Nanotechnology and Its Governance

This book charts the development of nanotechnology in relation to society from the early years of the twenty-first century. It offers a sustained analysis of the life of nanotechnology, from the laboratory to society, from scientific promises to societal governance, and attempts to modulate developments.

Arie Rip is Professor Emeritus of Philosophy of Science and Technology in the School of Management and Governance of the University of Twente, the Netherlands.

History and Philosophy of Technoscience
Series Editor: Alfred Nordmann

For more information about this series, please visit: https://www.routledge.com/History-and-Philosophy-of-Technoscience/book-series/TECHNO

Nanotechnology and Its Governance

Arie Rip

Routledge
Taylor & Francis Group

LONDON AND NEW YORK

First published 2020
by Routledge
2 Park Square, Milton Park, Abingdon, Oxon OX14 4RN

and by Routledge
605 Third Avenue, New York, NY 10017

First issued in paperback 2021

Routledge is an imprint of the Taylor & Francis Group, an informa business

British Library Cataloguing-in-Publication Data
A catalogue record for this book is available from the British Library

Library of Congress Cataloging-in-Publication Data
Names: Rip, Arie, 1941– author.
Title: Nanotechnology and its governance / Arie Rip.
Description: Abingdon, Oxon ; New York, NY : Routledge, 2019. | Includes
    bibliographical references and index.
Identifiers: LCCN 2019006581 | ISBN 9781138610538 (hbk) |
    ISBN 9780429465734 (ebk) | ISBN 9780429879517 (epub) |
    ISBN 9780429879524 (adobe) | ISBN 9780429879500 (mobi)
Subjects: LCSH: Nanotechnology. | Technology and state.
Classification: LCC T174.7 .R57 2019 | DDC 620/.5—dc23
LC record available at https://lccn.loc.gov/2019006581

ISBN 13: 978-0-367-78620-5 (pbk)
ISBN 13: 978-1-138-61053-8 (hbk)

Typeset in Times New Roman
by Apex CoVantage, LLC

Contents

1 Introduction

The life and times of nanotechnology and its governance

Arie Rip

When selecting papers for this collection, I was conscious of a deep sense of our late-modern society and the challenge of its attempt to handle technology, in particular a newly emerging technology like nanotechnology, with its combination of promises and concerns. This sense has coloured my work since my turn from chemistry (and philosophy) to science, technology, and society studies. I have tried to articulate elements of a diagnosis in my work on the danger culture of our industrial society, taking the handling of chemicals as my main example (Rip 1991): like the danger culture of miners and mountaineers, where rules are created and maintained which are functional to reduce danger, but are also taboos (cf. Mary Douglas 1966). This allows them (and us) to accept to continue living a dangerous life, up to life in industrial society.

For the case of nanotechnology, I had followed what was happening over time, at first occasionally because of my interest in how expectations are voiced about new technologies (see also Van Lente 1993), and more systematically since the early 2000s. I had seen the discussion about risks, e.g. of free nanoparticles, emerge. The issue of governance was (and is) much broader than regulation of risks of nanotechnology, however. It is about agenda building and implications for action (cf. Chapter 5). And it is about perceptions and actions, not in the least of nanotechnologists themselves (cf. Chapter 4). In other words, the governance of nanotechnology is to a large extent *de facto* governance, while occasionally punctuated by debates about government regulation (see Chapter 7; see also Kearnes and Rip 2009).

As Scharpf (1997: 204) phrases it (quoting Elinor Ostrom),

> Much effective policy is produced not in the standard constitutional mode of hierarchical state power, legitimated by majoritarian accountability, but rather in associations and through collective negotiations with or among organizations that are formally part of the self-organization of civil society rather than of the policy-making system of the state.

In the broadest sense of the concept of governance, all structuring of action and interaction that has some authority and/or legitimacy counts as governance. Authors on governance such as Van Kersbergen and Van Waarden (2004) and

Kooiman (2003) recognize this, even if they do not thematize it. Governance arrangements may be designed to serve a purpose, but can also just emerge and become forceful when institutionalized. The same move is visible in Voß et al. (2006: 8) where they argue that governance refers to 'the characteristic processes by which society defines and handles its problems. In this general sense, governance is about the self-steering of society'.[1]

In the sense outlined here, *de facto* governance is studied by, or should be studied by, sociology, and what I like to call 'macro-anthropology'. A macro-anthropology of nanotechnology allows us to address the richness of nanotechnology and its de facto governance.[2] This book shows possibilities, but does not give a comprehensive analysis or an overview. Up to its neglecting to thematize features like cultural aspects, that might be missed in a too exclusive focus on the technicalities of governance.

One interesting symptom of *de facto* governance, and thus a topic for macro-anthropology, is the way the announcement of support for nanotechnology by the US Clinton Administration early in 2000 created a socio-political entity 'nano-technology', clinching a variety of developments into a concerted effort (for a history of the specific dynamics, see McCray 2005), and with repercussions in the US and worldwide (see Chapter 2, see also Van der Most 2009). The life and times of nanotechnology started in 2000, a macro-anthropologist would argue. As our analysis shows, it coincided with the need of policy makers and administrators to show they could be strategic (cf. Rip 2002)

Although generally recognized as important for the field of nanotechnology, this is not the common 'origin myth' of nanotechnology, however.[3] By now, the regular story picks up on Richard P. Feynman's Christmas dinner speech for the American Physical Society in 1959, in which he unfolded his vision of interactions at the nano-scale, i.e. orders of magnitude of 1 to 100 nanometre (10^{-9} m), using the engaging title 'There's lots of room at the bottom' (Feynman 1960). The regular story has all the trappings of a myth, a key feature being its retrospective construction.[4] As Christopher Toumey (2008) has shown, there was little or no reference to Feynman's dinner speech until the early 2000s, when a community of nanotechnologists had started to form, incited by the new opportunities to get funding, linked to a general interest in the field fuelled by Eric Drexler's strong claims about so-called molecular manufacturing (see Chapter 5).

This may be seen as only of anecdotal interest, but it reinforces my general point about the importance of an anthropological approach, also for newly emerging technologies like nanotechnology.

In a sense, I have been fortunate to have access to nanotechnologists and nanotechnology policy makers and administrators from a relatively early stage. In 2003, given my record of work in chemistry and society, and in science, technology, and society more generally, I was invited to participate in the new Dutch national R&D programme on nanotechnology, eventually called NanoNed (see Rip and Van Lente 2013). We were recognizable as 'visiting strangers', but also legitimated in our actions because we were invited by the board of the national programme consortium, and thought to be part of the attempt to show politicians and wider publics that nanotechnology was being developed responsibly.

Because of NanoNed, I had legitimation to participate in meetings and informal discussions at the European Commission, and participated (together with PhD students and post-docs) in two European Networks of Excellence, Nano2Life and Nanobioraise, and also globally, in particular in the meetings of the International Dialogue on Responsible Development of Nanoscience and Nanotechnology (see Chapter 7). I was able to move about, as an anthropologist would do, up to interacting with the 'natives' and having them confide in me (see for a background argument about the role of 'moving about', Rip 2000).

A specific entrance point was our work on Constructive Technology Assessment, a version of TA which involves itself in the actual construction of new technology, attempting to broaden the aspects and actors taken into account. This includes a methodology of 'insertion' (cf. Chapter 8), essentially an ethnographic approach but includes a diagnosis of what was happening in and around nanotechnology. However, the present collection of articles is not about our work on Constructive Technology Assessment (see for Constructive TA, in addition to Chapter 8, Rip 1994; Schot and Rip 1997; Te Kulve and Rip 2008; Robinson 2009; Parandian 2012; Parandian and Rip 2013).

In this collection of published articles, I tell the story of nanotechnology as a socio-political entity and its de facto governance. As for many newly emerging technologies, the storyline starts with big promises – which then evoke big concerns. There is hype from both sides, but then life as usual reasserts itself, increasingly so from 2006 onwards. But there will be residues, i.e. insights and arrangements that continue and shape what will happen.

I start with an analysis of the nature of the promise. Not in the sense of the expectations per se, but about how they function to create a protected space for the new field to work on its own challenges, while being legitimated to do so by the promises (Chapter 2). This is actually a feature of present day technosciences, and of the move towards 'strategic science' in general (Rip 2002). The article adds another example of a promising new field, sustainability research, with different dynamics, but its main interest for this collection is the compact history of nanotechnology.

I then move towards a more detailed look at how this works out in practice, up to 'waiting games' of nanotechnology enactors, knowing about the promise but reluctant to invest in an indeterminate future (Chapter 3). I continue a micro- and meso-anthropological focus when looking at nanotechnologists working with folk theories about what is happening (Chapter 4). Particularly important was their concern that nanotechnology should not fall victim to the fate of genetically modified approaches in biotechnology,[5] which led them to take initiatives at interaction with publics at an early stage. One effect was a gradual shift in the view of roles and responsibilities of nano-enactors, with the attendant popularity of the notion of a 'responsible development' of nanotechnology, visible, among other things, in various proposals to have an ethical code for nanoscience and nanotechnologies.

We offered a diagnosis of what was happening at the time (Chapter 6), and I have expanded it in more general terms as a changing division of moral labour (Rip 2017).[6]

There is a distinct shift from the focus on more or less speculative promises and concerns in the first five years or so, to more concrete, and often more subdued claims. For the debates about risk and regulation, we have located the transition around 2006 (see Chapter 5). This is not to imply that problems were then solved, but the dimensions along which to address them appeared to be clearer. See also the case of the Danish construction industry which in their marketing strategies moved from being 'loud' (outspoken) in the early 2000s about their nano-activities to being 'silent' towards the end of the decade (Andersen 2011).

Not much later, the public interest moved to other issues, for example synthetic biology now drawing a lot of attention, also from social scientists and humanities scholars (compare Downs 1972 on the issue-attention cycle).

This is where I could conclude my story about nanotechnology, but the story is not just about the vagaries of issue-attention cycles and newly emerging technologies.[7] Nor was it a matter of 'we' having got nanotechnology 'right' this time. There are deeper issues, not necessarily specific to nanotechnology, and which a macro-anthropology should include. In particular there is the cultural image of a new technology coming in from the outside, and to be contained or tamed – or domesticated, to use an anthropological turn-of-phrase introduced by Roger Silverstone (1994). 'The metaphor of "domestication" came from the taming of wild animals, but was then applied to describing the processes involved in "domesticating ICTs" when bringing them in the home' (Haddon 2007: 26). I note in passing that these processes of domestication occur on location and cannot simply be shaped by nanotechnologists wanting 'to do it right'.[8]

New and emerging science and technology are positioned as being introduced into society, coming in from the outside as it were (Swierstra and Rip 2007). Thus, the technology is seen as having agency by itself, while society (societal actors) must come to terms with it. Enactors (the developers and promotors of nanotechnology) can then speak in the name of the exogenous technology, and feel justified in imposing their view of progress on society – up to blaming obstacles to such progress on phobias: nano-phobia (cf. Chapter 4 on enactor perspective).[9]

Outsiders to technology, especially when they are critical, can easily fall into the trap of attributing agency to exogenous technology. Sarah Franklin (2006, p. 87) illustrates this when she says,

> This view [of Fukuyama and Habermaß] of genetic manipulation as *a force unto itself*, hostile to social order and integration. . . . Here . . . 'biotechnology' is attributed a sinister agency.

Thus, outsiders picturing new technology as an independent force become part of an unholy alliance with insiders, perpetuating the myth of exogenous technology.[10]

In other words, the notion of domestication is not an innocent descriptive concept. It is, intentionally or unintentionally, part of the overall diagnosis, with implications for how we look at social order. One way to bring this out is to reflect on the irony of my story's 'happy' end: 'Don't worry, nanotechnology will be domesticated in the end'. I'm not just being flippant here. It raises the question if it is just a matter of decline of public interest, as the common interpretation of Downs (1972) would have it (cf. note 7). There might also be learning, about how to handle a new technology, and not just by deciding to 'do better this time'.

One can also worry about the possible costs involved, including opportunity costs.[11] I have considered learning (and costs) in general terms in my 'philosophy' of TA, intended to broaden the policy analysis approach to TA and a simple focus on cost-benefit views. The goal of TA as a societal function should be to anticipate to reduce the human cost of learning by trial and error about new technology (Rip 2012, 2015).

This philosophy is informed by a Science, Technology, and Society perspective: new science and technology emerge in an existing socio-technical landscape and are shaped by it while they also transform it. Assessment and governance of new science and technology should not look at them in isolation, and focus on consequentialist checking of impacts, attributing utilities, and then taking measures. A socio-technical landscape enables and constrains actions and interactions, and in that way functions as part of a *de facto* constitution of our society.

The basic question is about what sort of world do we, and can we, live in. In particular, what sort of governance arrangements can we set up and maintain with respect to new and emerging science and technologies? There is no simple answer, but there are entrance points – if one forgets about centralized governance and its illusion of control. Instead, modulation of ongoing dynamics is possible (Rip 2006, 2010).

It is for this reason that I felt it necessary to include an article on the overall approach of Constructive Technology Assessment in this collection (Chapter 8). Constructive TA does not start with regulation, but with improving de facto governance, in particular through so-called bridging events with various relevant actors, supported by sociotechnical scenarios which bring out tensions and dilemmas. There is no assurance that the goal of 'a better technology in a better society' (Schot and Rip 1997) will indeed be reached, but at least the level of reflexivity will be enhanced (Parandian 2012).

The methodology of 'insertion' that we describe in Chapter 8 (and actually used more generally when we were moving about in the nano-world and more widely) is indebted to anthropology; and may add to it as a soft version of critical anthropology.

Thus, this collection of articles has a broader message, in addition to its story about the life and times of nanotechnology and its governance. And I phrased it as part of general argument about the practice of social research (Rip 2000): *following actors in an unfolding story allows you to capture a storyline, and tell it to others*. That's what I have done, and I am sharing the story with you.

Notes

1 They develop this further: 'governance is understood as the result of interaction of many actors who have their own particular problems, define goals and follow strategies to achieve them. Governance therefore also involves conflicting interests and struggle for dominance. From these interactions, however, certain patterns emerge, including national policy styles, regulatory arrangements, forms of organisational management and the structures of sectoral networks. These patterns display the specific ways in which social entities are governed. They comprise processes by which collective processes are defined and analysed, processes by which goals and assessments of solutions are formulated and processes in which action strategies are coordinated. . . . As such, governance takes place in coupled and overlapping arenas of interaction: in research and science, public discourse, companies, policy making and other venues' (ibid.).

2 My approach is novel, but there are precursors and inspiring work, like Hans Glimell (2004), in a critically engaged vein. Also, assuming a more distantiated intellectual analysis, Colin Milburn (2008, 2015) on nano in popular culture. See also the Harthorn and Mohr edited volume (2012).

3 There is occasional reference to the first definition of nanotechnology by Taniguchi in 1974, often with a footnote that Taniguchi was writing about another concept, precision machining (McCray 2005: 196). Thus, it is not a story about origins, but about giving due reference.

4 There's more to say about Feynman's speech including the overall West Coast culture in which science fiction was an integral element. Colin Milburn has shown the influence of Robert Heinlein's short SF novel Waldo on Feynman's thinking at the time. (Milburn 2004)

5 Explicitly, as when Vicky Colvin referred to the 'wow-yuck' curve of GM biotech that had to be avoided by including ethical, legal, and social implications of nanotechnology (Ch4), or implicitly as in the conviction that 'this time, we'll do it right from the very beginning', as it pervaded some of the first big conferences organized under the auspices of the NNI (Hans Glimell, personal communication 2004, see also his reference to 'making everything right from the start' (Glimell 2004: 232)). It left traces (as in the introduction of Roco and Bainbridge 2001, and in the way a collaborative project of Dupont Company and NGO Environmental Defense Fund on risks of nanotechnology was announced as 'getting nanotechnology right', see Krupp and Holliday 2005).

One important institutional effect of this prevailing conviction that 'we' have to get nanotechnology 'right' was the call, in 2004, for an NSF-funded Center for Nanotechnology in Society. It eventually resulted in the establishment, in 2005, of two such centres (actually, networks linking different institutions), one centred on Arizona State University and one centred on the University of California Santa Barbara.

6 This is already visible in everyday life, where narratives of praise and blame are key feature of social life, often somewhat institutionalized as a repertoire. This has governance implications. We have analysed the debate on labelling of nano-enabled products, cosmetics being one example (Throne-Holst and Rip 2011), where preferred measures can be distinguished as to whom a final responsibility is relegated: a government specifying what is generally recognized as safe; or the consumer as buyer of the product making individual decision; or the producer facing the possibility of liability cases because of an implied warrant.

7 Actually, the original article of Downs (1972), taking the rise and then fall of ecological concerns as its topic, is much more sophisticated than the later use of the issue-attention cycle, with its focus on media attention, would lead one to think. It creates a five-phase diagnosis, which is more substantial, and is surprisingly applicable to the story of nanotechnology and its governance. It starts with a latent (or pre-problem) phase where issues are not noticed; then there is a phase of 'alarmed discovery and euphoric enthousiasm' (about addressing the problem/issues); in a third phase, there

is 'realising the cost of significant progress', followed by a fourth phase of 'gradual decline of intense public interest'. The final phase Downs calls the 'post-problem stage'.

8 The location need not be the home, but can be the society as a whole. See Lie and Sørensen (1996).

9 The original phrase was used in the 1970s to characterize the concern of chemists about the public image of chemistry as chemophobia-phobia, leading them to see fear of chemistry where it was not (Rip 1986).

10 There is a deeper layer to this argument, explored by Alfred Nordmann (2006), starting with the recognition that nano-scale processes are essentially inaccessible, and can only be traced in terms of their effects. Thus, they are examples of unknowable things-as-such, Kant's *Ding-an-Sich*, or noumenal technology. For Nordmann, this shows a form of technology that is not controlling nature, but 'as uncanny as brute, uncomprehended nature itself' (Nordmann 2006: 49). Nanotechnology allows pervasive technical interventions which 'change the things-in-themselves, the world not as we know it but where we rely on it unknowingly' (Nordmann 2006: 67).

11 One example is the tendency to focus on new technology offering upstream solutions of what are essentially downstream problems – which then do not get the attention they deserve (Rip 2009).

Bibliography

Andersen, Maj Munch (2011), 'Silent Innovation: Corporate Strategizing in Early Nano-technology'. *Journal of Technology Transfer* 36(6), pp. 680–696.

Douglas, Mary (1966), *Purity and Danger: An Analysis of Concepts of Pollution and Taboo*. London: Routledge & Kegan Paul.

Downs, Anthony (1972), 'Up and Down With Ecology: The Issue-Attention Cycle'. *The Public Interest* 28, pp. 38–50.

Feynman, Richard P. (1960), 'There's Plenty of Room at the Bottom: An Invitation to Enter a New Field of Physics'. *Engineering and Science* 23(5), pp. 22–26. Made available on Internet, www.zyvex.com/nanotech/feynman.html (accessed 16 October 2003).

Franklin, Sarah (2006), 'Better by Design?'. In Paul Miller and James Wilsdon (eds.), *Better Humans? The Politics of Human Enhancement and Life Extension*. London: Demos, pp. 86–94.

Glimell, Hans (2004), 'Grand Visions and Lilliput Politics: Staging the Exploration of the "Endless Frontier"'. In Davis Baird, Alfred Nordmann, and Joachim Schummer (eds.), *Discovering the Nanoscale*. Amsterdam: IOS Press, pp. 231–246.

Haddon, Leslie (2007), 'Roger Silverstone's Legacies: Domestication'. *New Media & Society* 9(1), pp. 25–32.

Harthorn, Barbara Herr, and John W. Mohr (eds.) (2012), *The Social Life of Nanotechnology*. New York: Routledge.

Kearnes, Matthew, and Arie Rip (2009), 'The Emergent Governance Landscape of Nanotechnology'. In Stefan Gammel, Andreas Lösch, and Alfred Nordmann (eds.), *Jenseits von Regulierung: Zum politischen Umgang mit der Nanotechnologie*. Heidelberg: Akademische Verlagsgesellschaft, pp. 97–121.

Kooiman, Johannes (2003), *Governing as Governance*. London: Sage Publications.

Krupp, Fred, and C. Holliday (2005), 'Let's get Nanotech Right'. *Wall Street Journal*, Tuesday, June 14, 2005, Management Supplement, B2.

Lie, Merete, and Knut Sørensen (1996), *Making Technologies Our Own? Domesticating Technology into Everyday Life*. Oslo: Scandinavian University Press.

McCray, W. Patrick (2005), 'Will Small Be Beautiful? Making Policies for Our Nanotech Future'. *History and Technology* 21(2), June, pp. 177–203.

Milburn, Colin (2004), 'Nanotechnology in the Age of Posthuman Engineering: Science Fiction as Science'. In N.K. Hayles (ed.), *Nanoculture: Implications of the New Technoscience*. Bristol, UK: Intellect Books, pp. 109–129 and 217–223. Reprinted from *Configurations* 10(2), Spring 2002, pp. 261–296.

Milburn, Colin (2008), *Nanovision: Engineering the Future*. Durham, NC: Duke University Press.

Milburn, Colin (2015), *Mondo Nano: Fun and Games in the World of Digital Matter*. Durham, NC: Duke University Press.

Nordmann, Alfred (2006), 'Noumenal Technology: Reflections on the Incredible Tininess of Nano'. In Joachim Schummer and Davis Baird (eds.), *Nanotechnology Challenges: Implications for Philosophy, Ethics and Society*. Singapore: World Scientific Publishing Co, pp. 49–72.

Parandian, Alireza (2012), *Constructive TA of Newly Emerging Technologies Stimulating Learning by Anticipation Through Bridging Events* (PhD Thesis, TU Delft, defended 12 March 2012).

Parandian, Alireza, and Arie Rip (2013), 'Scenarios to Explore the Futures of the Emerging Technology of Organic and Large Area Electronics'. *European Journal of Futures Research* 1(1), published online 13 July 2013, doi:10.1007/s40309-013-0009-2.

Rip, Arie (1986), 'Legitimations of Science in a Changing World'. In Theo Bungarten (Hrsg.), *Wissenschaftssprache und Gesellschaft*. Hamburg: Edition Akademion, pp. 133–148.

Rip, Arie (1991), 'The Danger Culture of Industrial Society'. In Roger E. Kasperson and Pieter Jan M. Stallen (eds.), *Communicating Risks to the Public: International Perspectives*. Dordrecht: Kluwer Academic, pp. 345–365.

Rip, Arie (1994), 'Science & Technology Studies and Constructive Technology Assessment'. *EASST Newsletter* 13(3), ISSN 0254 9603, September 1994, pp. 11–16. Keynote Speech to EASST Conference, Budapest, 28–31 August 1994. With Comments by John Ziman, Les Levidov, and Andrew Barry.

Rip, Arie (2000), 'Following Actors – Then What?'. Invited Paper, *Seminar Neuere Ansätze und Methoden in der Wissenschafts- und Technikforschung*. Technische Universität Darmstadt, 5 May 2000.

Rip, Arie (2002), 'Regional Innovation Systems and the Advent of Strategic Science'. *Journal of Technology Transfer* 27, pp. 123–131.

Rip, Arie (2006), 'The Tension Between Fiction and Precaution in Nanotechnology'. In Elizabeth Fisher, Judith Jones, and Rene von Schomberg (eds.), *Implementing the Precautionary Principle: Perspectives and Prospects*. Edward Elgar, pp. 423–448.

Rip, Arie (2009), 'Governance of New and Emerging Science and Technology'. In Peter Healey and Steve Rayner (eds.), *Unnatural Selection: The Challenges of Engineering Tomorrow's People*. London and Sterling, VA: Earthscan, pp. 209–214.

Rip, Arie (2010), 'Processes of Technological Innovation in Context – And Their Modulation'. In Chris Steyart and Bart van Looy (eds.), *Relational Practices, Participative Organizing*. Bingley, UK: Emerald, Advanced Series in Management, pp. 199–217.

Rip, Arie (2012), 'Futures of Technology Assessment'. In M. Decker, A. Grunwald, and M. Knapp (Hrsg.), *Der Systemblick auf Innovation – Technikfolgenabschätzung in der Technikgestaltung*. Berlin: Edition Sigma Verlag, pp. 29–39.

Rip, Arie (2015), 'Technology Assessment'. In James D. Wright (ed.), *International Encyclopedia of the Social & Behavioral Sciences*, 2nd edition. Oxford: Elsevier, pp. 125–128.

Rip, Arie (2017), 'Division of Moral Labour as an Element in the Governance of Emerging Technologies'. In Diana M. Bowman, Elen Stokes, and Arie Rip (eds.), *Embedding New Technologies into Society: A Regulatory, Ethical and Societal Perspective*. Singapore: Pan Stanford, November, pp. 115–129.

Rip, Arie, and Haico te Kulve (2008), 'Constructive Technology Assessment and Socio-technical Scenarios'. In Erik Fisher, Cynthia Selin, and Jameson M. Wetmore (eds.), *The Yearbook of Nanotechnology in Society, Volume I: Presenting Futures*. Berlin etc: Springer, pp. 49–70.

Rip, Arie, and Harro van Lente (2013), 'Bridging the Gap Between Innovation and ELSA: The TA Program in the Dutch Nano-R&D Program NanoNed'. *Nanoethics* 7(1), April, pp. 7–16.

Robinson, Douglas K.R. (2009), 'Ço-Evolutionary Scenarios: An Application to Prospecting Futures of the Responsible Development of Nanotechnology'. *Technological Forecasting and Social Change* 76, pp. 1222–1239.

Roco, Mihael, and William S. Bainbridge (eds.) (2001), *Societal Implications of Nanoscience and Nanotechnology*. Arlington, VA: National Science Foundation.

Scharpf, Fritz W. (1997), *Games Real People Play: Actor-Centred Institutionalism in Policy Research*. Boulder, CO: Westview Press.

Schot, Johan, and Arie Rip (1997), 'The Past and Future of Constructive Technology Assessment'. *Technological Forecasting and Social Change* 54, pp. 251–268.

Silverstone, Roger (1994), *Television and Everyday Life*. London: Routledge.

Swierstra Tsjalling, and Arie Rip (2007), 'Nano-Ethics as NEST-Ethics: Patterns of Moral Argumentation about New and Emerging Science and Technology'. *NanoEthics* 1, pp. 3–20.

Throne-Holst, Harald, and Arie Rip (2011), 'Complexities of Labeling of Nano-Products on the Consumer Markets'. *European Journal of Law and Technology* 2(3), pp. 1–12.

Toumey, Christopher (2008), 'Reading Feynman into Nanotechnology: A Text for a New Science'. *Techne* 12(3), pp. 133–168.

Van der Most, Frank (2009), *Research Councils Facing New Science and Technology: The Case of Nanotechnology in Finland, the Netherlands, Norway, and Switzerland* (PhD Thesis, University of Twente, defended on 13 November 2009).

Van Kersbergen, Kees, and Frans Van Waarden (2004), 'Governance as a Bridge Between Disciplines: Cross-Disciplinary Inspiration Regarding Shifts in Governance and Problems of Governability, Accountability and Legitimacy'. *European Journal of Political Research* 43, pp. 143–171.

Van Lente, Harro (1993), *Promising Technology: The Dynamics of Expectations in Technological Developments* (PhD Thesis, University of Twente, Enschede, defended 25 November 1993).

Voß, Jan-Peter, Dierck Bauknegt, and René Kemp (eds.) (2006), *Reflexive Governance for Sustainable Development*. Cheltenham: Edward Elgar.

2 Umbrella terms as mediators in the governance of emerging science and technology

Arie Rip and Jan-Peter Voß

Published as

Arie Rip and Jan-Peter Voß, Umbrella terms as mediators in the governance of emerging science and technology. *Science, Technology, and Innovation Studies*, 9 (2013) 39–59.

Introduction

It is intriguing how new fields of science emerge, such as nanotechnology or sustainability research, in recent decades with names that do not only indicate a field of research but also carry promises of major industrial transformation (in the case of nanotechnology) or claim to address daunting problems (in the case of sustainability research). What we see here is the intersection of two developments: a longer tradition of emerging new fields labelled to entail a scientific promise, as with physical chemistry in the late nineteenth century and colloid science in the early twentieth century, and a recent transformation of science in the direction of strategic science (Rip 2002), where long-term relevance to societal problems, hence a societal promise, is an integral part of how the science is done. The intersection of the two developments is visible, if we take a look at how labels like 'nanotechnology' or 'sustainability research' are used and what they do to shape and hold together certain patterns in the de facto governance of science. In the light of this function, we propose that the labels be called umbrella terms.

Our argument in this paper is that, in studying the mechanisms of governance that shape scientific development de facto, it is worthwhile taking a closer look at the organizing qualities of particular terms that can work to connect and mediate a variety of activities and concerns across different fields of science, science policy and society – even without any explicit frameworks structuring those relations de jure. They link up and translate discursively patterned practices. Umbrella terms start out as a fragile proposal by which a variety of research areas and directions can be linked up with one another, and, in a sense, 'covered' (which is where the metaphor of an umbrella comes in), with a view to relating them, as a whole, with certain societal concerns and policy issues. In this way they provide a semantic reference for negotiating certain packages of scientific search practices with societal and political concerns. Over time, umbrella terms and the packages they hold together may stabilize and become reinforced with research infrastructures and institutionalization of funding schemes.

This phenomenon of umbrella terms as mediators that enable the creation and functioning of packages of scientific research and policy and societal relevance indicates a new way in which science is being governed – de facto. This deserves to be explored, and not just in science policy studies, with their occupational bias of prioritizing policy. Science and technology studies (STS) have to contribute because of their tradition of studying dynamics of scientific developments in context. Such a study of governance of science is a relatively new venture for STS,[1] particularly when we consider how the study of umbrella terms, their emergence and possible stabilization, even when carried merely in the form of a scholarly study, will have implications for the governance of science and the role played in it by STS scholars. The attention paid to a specific umbrella term will reinforce its status, even when the study actually deconstructs the ongoing processes.[2] This is unavoidable. It is also an indication that conducting STS in the real world requires further reflexivity.

We will explore the nature of the intersection of emerging scientific fields and strategic science, this being the location of the phenomenon of umbrella terms, in two steps. First, we will characterize the phenomenon of umbrella terms and locate it in present day science in its respective contexts. Second, we will present two case studies with interesting differences, namely nanotechnology and sustainability research. Nanotechnology has become institutionalized as a field bearing this label, sustainability research has not, or at best has only done so to a partial extent, because different labels are competing to configure the science/policy link in particular ways. Furthermore, nanotechnology is about opportunities and promises opened up by techno-scientific developments (with open and flexible linkages to societal and policy promises), while sustainability research (and its variants) is an attempt to mobilize and position different scientific developments in relation to a socio-politically constructed global problem. Both are instances of the phenomenon of umbrella terms and how these function, broadly speaking, as mediators between science and science policy. In the concluding section we will reflect on the type of governance we can observe here, and also ask what our own role is in studying these developments.

Umbrella terms marking the intersection between strategic science and emerging scientific fields

Over the last three decades, the practices of scientific research, the institutions of science, and their concrete contexts have all been changing, and there has been recognition of, and reflection on, these changes. There have been attempts to diagnose these changes, or certain aspects of them (Funtowicz and Ravetz 1993; Ziman 1994; Gibbons et al. 1994; Nowotny et al. 2001; Etzkowitz and Leydesdorff 2000; Bonaccorsi 2008; see also Bonaccorsi 2010; Lave et al. 2010). What is clear is that there is a general movement towards re-contextualization of science in ongoing processes in wider areas of society (Nowotny et al. 2001; Markus et al. 2009), and that a new regime of Strategic Science has emerged after the opening up of the earlier regime in place since the Second World War (sometimes called Science, The Endless Frontier, after the title of the influential report of Vannevar Bush to the US President in 1945 (Bush 1945)). The opening up of this regime is already indicated in the influential 1971 Brooks Report to the OECD (OECD 1971), in which closer and more differentiated links between science and society were advocated, in contrast to the earlier regime where 'science' is considered to be a unified whole. The next phase is indicated by the introduction of the notion of strategic research, linking basic research to societal problems and challenges. Irvine and Martin's (1984) characterization of strategic research captures the nature of this link, indicating a new division of labour between the quest for excellence and for relevance:

> Strategic research is
>
> • basic research carried out with the expectation that it will produce a broad base of knowledge
> • likely to form the background to the solution of recognized current or future practical problems

The link is formulated in terms of expectations, but there are also new practices as when research funding agencies started creating strategic research programmes,[3] and centres for excellent and relevant research were established inside or outside universities since the 1980s, their continuing viability deriving from the emergence of markets of strategic research (Rip 2002). Also, priority setting became linked to foresight exercises. Such developments can be seen as creating institutionalized 'trading zones' between science and societal issues and their spokespersons.[4] Thus, there are reasons to speak of a new regime, a regime of Strategic Science. There are other developments as well, not directly related to strategic research, but compatible with it: the rechannelling of resources for scientific research through competitive project funding compared to block funding for universities and public research institutes, and the establishment of new audit and evaluation procedures.

In the 'trading zones' one sees packaging of social questions, opportunities, and scientific developments, which can be 'sold' to various audiences and which are often labelled so as to carry rhetorical force. An early example is the 'War on Cancer' programme in the US in the 1980s (Rettig 1977). A recent example of such packaging is the discourse of 'Grand Challenges' in Europe and elsewhere (cf. EU: Lund Declaration, Horizon 2020). The way that the UK Research Councils have defined and outlined ten Grand Challenges (RCUK 2009) is illustrative of this, some in a technology-push or scientific-opportunity-driven mode, others in a society-pull or social-problem-driven mode. In Box 2.1 we quote two of them in some detail, which will also allow us to refer to them in our further discussion.

Box 2.1   Two 'Grand Challenges' defined by research councils UK

| *NanoScience through Engineering to Application* | *Ageing: life-long health and wellbeing* |

Nanotechnologies can revolutionize society. They offer the potential of disruptive step changes in electronic materials, optics, computing, and in the application of physical and chemical understanding (in combination with biology) to generate novel and innovative self-assembled systems. The field is maturing rapidly, with a trend towards ever more complex, integrated nanosystems and structures. It is estimated that products incorporating nanotechnology will contribute US$1 trillion to the global economy by 2015, and that the UK has a 10 percent share of the current market. To focus the UK research effort we will work through a series of Grand Challenges. These will be developed in conjunction with researchers and users in areas of societal importance such as energy, environmental remediation, the digital economy, and healthcare. An interdisciplinary, stage-gate approach spanning basic research through to application will be used. This will include studies on risk governance, economics, and social implications.

There is an unprecedented demographic change underway in the UK with the proportion of young people declining whilst that of older people is increasing. By 2051, 40 percent of the population will be over 50 and one in four over 65. There are considerable benefits to the UK of having an active and healthy older population with potential economic, social, and health gains associated with healthy ageing and reducing dependency in later life. Ageing research is a long-standing priority area for the Research Councils. The Research Councils will develop a new interdisciplinary initiative (£486 million, investment over the CSR period involving all seven Research Councils) which will provide substantial longer-term funding for new interdisciplinary centres targeting themes of healthy ageing and factors over the whole life course that may be major determinants of health and wellbeing in later life. Centres will be focussed on specific research themes drawing on the interdisciplinary strengths of the Research Councils, such as Quality of Life, Physical Frailty, and Ageing Brain.

In these examples, a short phrase summarizes the thrust of the Grand Challenge. For the second Grand Challenge, the problem is often denoted as 'the ageing society', a label that allows easy reference to a set of complex interrelated issues, while at the same time black boxing them to some extent. Reference to 'the age-ing society' then becomes a justification to speak of 'ageing research' rather than more disciplinary-oriented names like 'biogerontology' (Miller 2009). The label 'ageing research' can become a package in its own right, referring to assorted research with a shared relevance to 'the ageing society'. This fits the notion of strategic research, but is now positioned on the field level rather than as research projects. In the first Grand Challenge, a similar easy reference coupled with some black boxing occurs through the label 'nanotechnology', as in the opening sentence (where the plural is used). The reference is to a techno-scientific field that definitely already exists as a funding category. Even so, it covers a wide range of items,[5] and for that reason can already be called an umbrella term.

Packaging of new scientific approaches with the help of labels occurred in the history of science, for example 'physical chemistry' in the late nineteenth century (Dolby 1976) and molecular biology from the 1930s onward (Bartels 1984; Kohler 1976). An interesting further example is the rise of the notion of colloid science in the 1910s and 1920s, when the term was presented as indicating a fourth phase of matter (in addition to solid, fluid, and gaseous) and the key to understanding the nature of living matter – and thus worthy of support and further exploitation (Ede 2007). Here, the audience for what starts as an umbrella term (because its scope is still unclear) is a scientifically concerned audience, and non-scientific audiences that put various issues of relevance upfront are involved only at one remove.

This continues to occur, but by now policy and other societal audiences are important as well. This implies that there is not only struggle for recognition (and funding) of new fields within science, but also struggle for legitimacy and resources in direct interaction with policy communities and a variety of social groups who are looking for opportunities to endorse and fund interesting research programmes. For society, this means a field of opportunities. For science, it often means space for new interdisciplinary approaches. And the promise of opportunities encapsulated in the umbrella term provides a protected space for such new approaches. The broad base of knowledge to be created through basic research, likely to form the background to the solution of future problems (cf. Irvine and Martin's definition of strategic research), is held in place by an umbrella term.

The phenomena we describe here have been noted and conceptualized before, in particular by the Starnberg-Bielefeld Group in their work on the so-called finalization thesis. Their original ideas centred on the diagnosis that fields have to become mature before relevance considerations can productively be included in scientific agendas, including 'finalized' theory development. Their conceptualization is based on how scientific paradigms, in the sense of Kuhn (1970), evolve, while this is just one aspect of inter-organizational fields of research. Their case studies, e.g. on environmental research and cancer research, did show more complex dynamics, as well as the role of umbrella terms (Böhme et al. 1978; Van den Daele et al. 1979; see also Schäfer 1983; Rip 1997). What they did not consider was the phenomenon of translation zones and mediators, while this has now become a striking feature of science in our society. Umbrella terms have become mediators between the logics of scientific search and the logics of various societal and policy worlds, and are thus constitutive of new patterns of re-contextualized science and technology.

Umbrella terms and their dynamics

While an umbrella term is a part of discourse, its use in ongoing struggles (e.g. in building coalitions of scientists and policy actors) and its eventual wider acceptance in labelling organizations and programmes turns it into an institutional and practical reality. The inter-organizational field of research organizations, relevant government agencies, civil society organizations, and representatives from domains of application acquires coherence and stability through reference to the umbrella term.[6] Thus, it is important to understand how umbrella terms become forceful as mediators between science and science policy and society.

Let us start by identifying examples. We mentioned ageing research and nanotechnology already. An earlier (and less grandiose) example is membrane science and technology since the 1970s, where the promises created a space that was filled in by dedicated R&D and gradually realized functionalities (Van Lente and Rip 1998). There are other (sometimes partial) examples like synthetic biology, or geo-engineering, both of them definitely on the radar of science policy actors and funding agencies at the moment.

The umbrella terms can also start from the other side, when the entrance point is a newly articulated function to be fulfilled, by different scientific and technological developments. Examples are 'targeted drug delivery' and 'personalized medicine', or 'the information superhighway' of the early 1990s, promoted by Al Gore among others. Kornelia Konrad has shown the power of this umbrella term in how it led government agencies and city governments to invest in projects, and when these failed, attributed it to contingencies so that they would invest in further projects rather than reconsider the promise (Konrad 2004, 2006). Security studies are an example where a number of different fields merged, or at least collaborated, under this umbrella term to address topics high on the political agenda. A further example is how sustainability (and sustainable development) has become a powerful reference in discourses, also of science and science governance: as something like an ecologically extended version of the 'common good' it can be invoked as a metagrand challenge of world society. Relating activities and projects to it carries a diffuse but positive message, and can thus be used to mobilize resources. While sustainability itself is not an umbrella term in the specific sense of this paper, since it has not (yet) been established as a fixed term for talking about, supporting, and negotiating a bundle of concrete research activities, it is an entrance point to study ongoing attempts at creating a science of sustainability where various candidate terms circulate (e.g. global change research, earth system science, sustainability science). We will discuss this further in our second case study.

The umbrella terms are mediators through which scientific promises and definitions of public problems travel and get entangled in constructions of 'relevant science'.[7] Thus, the umbrella term is not just a word or a phrase in a discourse, it is also, eventually, a conduit through which specific scientific opportunities and promises interact with specific societal and policy goals and interests, thus providing for their mutual shaping.

We will consider the process of emergence and stabilization of umbrella terms, together with the interorganizational fields that are formed, a bit further. An umbrella term emerges in a specific constellation of discourse, activities, and incipient as well as more established institutionalization. This is not just a matter of scientists packaging promises. Science policy makers scan the horizon for productive fields that can be linked to a 'public interest' and occasionally they initiate or catalyze the formation of fields which they expect to be important and think matching societal support can be mobilized. Increasingly, large corporations and business associations, non-governmental organizations, and social movements also actively look out for research practices that promise relevance to their concerns and they engage with the framing of science-society relations.

There is a long tradition of opportunistic resource mobilization by scientists, as well as 'politicking' by spokespersons for science to assure symbolic resources for science (Rip 1990). A newly proposed umbrella term then is a way of packaging a proposal which offers an investment in scientific capacity: a 'sales proposal'. Some such sales proposals are more successful than others, and scientists will anticipate what is on the agenda in science policy and in society more generally, and adjust their proposal in terms of content, and definitely in terms of terminology. Intermediary actors like funding agencies, when they identify with science rather than policy, follow similar tactics (this is visible in the Grand Challenges discourse of the UK Research Councils). Further tactics of resource mobilization using visionary umbrella terms are visible to acquire funding on top of disciplinary funding structures, and/or to circumvent disciplinary funding structure for new, interdisciplinary research agendas. Occasionally, scientists refer to umbrella terms to offer their service directly to policy or society bypassing funding agencies.

If scientists offering their packages are seen as the supply side, the demand side consists of science policy makers and other sponsors of science wanting to provide funding (and other support) in an interesting and useful way. There will be reference to problem areas and societal challenges, to justify science policy and R&D programme budgets, which can lead to further articulation of such problems and challenges. In this sense, science policy makers can also be seen as brokers between scientific supply and societal demand.[8] The net effect is that a name or a phrase that works both ways, for policy as well as for science (or is made to work for both sides), helps to fill in the 'trading zones' and acquire a life of its own: an umbrella term is born.[9]

There will be expectations, policy declarations, strategy meetings, platforms, and other collective initiatives, programmes of research and new centres, dedicated intermediary organizations, etc. Further actors will join, which will involve some controversy and struggle over the definition of boundaries – what is in, what is out, what is at the centre, and what is only peripheral. This is an inter-organizational field, with epistemic components (one can speak of a new scientific field) as well as institutional, economic, and socio-political components linked to problems, challenges, and actual applications. There will also be public statements and media reporting, while scientists (and policy makers) will anticipate public reactions and civil society responses. Institutionalization then leads to specialized organizations, including education and training programmes. The umbrella term represents and helps to stabilize the inter-organizational field while it functions as a conduit between scientific activities and society.

Implicit in this stylized description of umbrella term dynamics in context is a further element, namely how 'demand' and 'supply' for scientific research can clinch through shared reference to an umbrella term and give the term force. In the case of nanotechnology, there was a very visible clinching event when the

US National Nanotechnology Initiative was announced in 2000. In the case of sustainable development, there is increasing interest from institutions and sponsors. Various local clinchings occur under labels which use modifications of the root term 'sustainable' and a recent attempt was made to bring a diversity of research networks and sponsors together for a global programme, 'Future Earth: Research for global sustainability', in which several of the currently advanced candidate terms appear in combination. This adds up to umbrella term dynamics, even without a single dominant clinching event that establishes a particular term as the reference for all ongoing attempts at configuring a science of sustainable development.[10]

Nanotechnology

Originally, the term 'nanotechnology' was used in an ad-hoc manner,[11] together with variants like 'nanoscale science' and 'nanoscale technologies'. Based on secondary literature and our own work and experience, we will trace its ascendancy as an umbrella term since the late 1990s, together with the emergence of an interorganizational field represented by and sustaining the force of the umbrella term. We will then explore its dynamics, and end with a brief diagnosis of the present situation.

In the 1990s, there was the visionary use of the term nanotechnology by Eric Drexler and his Foresight Institute (Drexler 1986), and the practical and somewhat ad hoc use in descriptions of funding programmes (Van der Most 2009). For many scientists, 'nanotechnology' was not important as a label. They were happy to do materials science or supra-molecular chemistry. The earlier funding programmes (since 1996 in Germany and Sweden, but already in 1994 in Switzerland) had specific topics, related to existing scientific fields and areas of application. The UK's earlier 'National Initiative on Nanotechnology' (since 1986) led by an alliance between the government Department of Trade and Industry and the National Physical Laboratories were similarly specific, even when using the general label. The two Nobel Prizes now listed as highlights in the development of nanotechnology, the 1986 Physics Prize for Scanning Tunneling Microscopy (first publication in 1980) and the 1996 Chemistry Prize for Fullerenes (or buckyballs; first publications in 1985), were seen as important in their own right, and only later became an argument for the importance of nanotechnology.[12] Thus, the term was available and was used, but not as an umbrella term.

The promise of research at the nanoscale was recognized,[13] but there were no grand visions, except for Drexler's programme of 'molecular manufacturing'. This programme was actively promoted by his Foresight Institute, established in 1986. It organized meetings and conferences, gathered followers and generated general interest.[14] Richard Smalley, who became critical of Drexler's programme, still acknowledged how he had been inspired by the vision and the meetings he attended (Regis 1996: 275–278).

The landscape changed with the US National Nanotechnology Initiative (NNI), announced in early 2000 by President Clinton. The NNI became a reference point for funding agencies and policy makers worldwide, and led to a 'funding race' (Rip 2011). It needed to justify itself in terms of promises, up to a third industrial revolution. Scientists started to refer more emphatically to nanotechnology in their funding proposals and presentations to the outside world. Research institutes and centres were renamed so as to include nanotechnology in their title (this was happening already, but NNI reinforced the trend). Journals appeared with nanotechnology (or the prefix nano) in the title such as NanoLetters (since 2000) and the Journal of Nanoscience and Nanotechnology (since 2000).[15] And meetings and platforms were organized to articulate strategies for nanotechnology R&D and innovation. The recent European Technology Platform Nanomedicine is a good example of such anticipatory coordination in terms of participants and topics (cf. also Rip 2012), while it is also clear that 'nanomedicine' is itself an umbrella term that covers very different developments, each with their own dynamics. Taking all this together, it is clear that the nascent inter-organizational field had solidified, together with its umbrella term 'nanotechnology'.

In recent years, nanotechnologists and policy makers have explicitly referred to nanotechnology as an umbrella term, although mostly to indicate the difficulties of defining nanotechnology and the variety of research areas and approaches under this heading. The European Commission started to use the plural: nanosciences and nanotechnologies. This is not just recognition of variety. It is a response to the homogenizing effect of using the term 'nanotechnology', and the problems this introduces in the societal and political debate about risks and regulations of 'nanotechnology'. The halo effect of the term 'nanotechnology' continues to be exploited, however, for example in the recent move to emphasize 'green nanotechnology' as the real promise.

Looking back, one can enquire into how the launching of the US NNI became the key event. There was fertile soil for what we called a clinching between supply and demand sides. By the late 1990s one sees attempts at stock taking by funding agencies in a number of countries, sometimes induced by leading scientists (Van der Most 2009). In the US, National Science Foundation's adviser for nanotechnology, Mihael Roco, organized a meeting in 1997 to bring disparate activities in nanoscience and nanotechnology together across different agencies. This led to the establishment of an Interagency Working Group which met throughout 1998 and worked out a vision for what ultimately became the NNI (McCray 2005: 185–186). What is striking is how NNI brought a large number of government ministries and agencies, not known for their willingness to collaborate in science funding and science policy, together in a concerted effort.

Roco acted as an institutional entrepreneur, but was also well embedded in the emerging world of nanoscience. He created and spread visions of nanotechnology, referring to nanotechnology in general rather than some specific field – in particular, visions of a third industrial revolution enabled by nanotechnology, and of nanotechnology, as the basis for converging technologies for human enhancement. The willingness of scientists and engineers to join in had to do with the prospect of increased funding, of course, but they could also share part of these visions about the promises of nanotechnology. At the 22 June 1999 meeting of the

House of Representatives' Committee on Science, nanoscientist Smalley could say, 'There is a growing sense in the scientific and technical community . . . that we are about to enter a golden new era' (McCray 2005: 187).[16]

The net effect outside the US was that countries started to consider nanotechnology a priority, or reinvigorated what they were doing already. Often, it was an alliance between scientists who wanted to mobilize resources by referring to the example of NNI, and a small but influential number of policy makers who wanted to buy into nanotechnology as a major new priority. As we noted already, a funding race emerged, where countries (and regions like European Union) compared their R&D expenditure on nanotechnology and argued that they should not lag behind. In spite of the reference to trillion-dollar markets and a third industrial revolution (originally offered to help justify NNI, and then adopted in policy documents all over the world), major innovations enabled by nanotechnology were slow to arrive. There was no innovation race in nanotechnology, and after the first round of enthusiasm (in the early and mid-2000s), venture capitalists started to withdraw.[17] The recent move to 'green' nanotechnology can be seen as a response: a way to recapture societal and investors' interest.

After the first enthusiasm and somewhat indiscriminate funding, which allowed scientists (now called 'nanoscientists') to pursue their interests, the late 2000s saw attempts from policy makers, partly because of pressure from political actors, to get some value for money, i.e. making sure that the research that was funded would be relevant. The RCUK Grand Challenge Nanotechnology emphasizing the route to applications (see Box 2.1) is one example.[18] In other words, 'nanotechnology' as a mediator between science, science policy, and society moved from primarily offering a protected space for scientists to also work in the other direction, thus ensuring the relevance of publicly funded research.

One can ask whether nanotechnology, i.e. nanosciences and nanotechnologies, is also becoming a new scientific field. There is productive interdisciplinarity, centring on the techno-scientific objects that are created and studied which then also create links to application/innovation.[19] Newly launched journals exploit the present visibility of nanotechnology (and some fail to survive, cf. Grienesen 2010). They create outlets for ongoing research, and thus contribute to the build-up and establishment of the field of nanosciences and nanotechnologies. The institutes and centres that use the nanotechnology label to present themselves are sites where the new scientific field can be nurtured. Such epistemic and institutional investments will remain in place when the nanotechnology hype has passed by.

Sustainability research

The term 'sustainable development' is a political construction which was drawn up in the context of the World Commission of Environment and Development (WCED 1987). The term marked an effort to unite concerns about the global environment with those about economic growth, and thus to overcome antagonistic positions of environmental movements and industry, as well as between North and South.[20] Since the 1990s we have also seen references to sustainable development or sustainability in relation to science. There are efforts to position research activity in relation with what appeared to become an overarching societal and political concern. Attempts were made to articulate 'sustainability science' or 'sustainability research' as a new epistemic programme. A variety of scientific initiatives and sponsors established themselves on the force of 'sustainability' as an ideograph.[21] We will report on these efforts by drawing on documents and websites, and on our own observations from doing research in sustainability related programmes. We give an account of how, in recent years, 'sustainability science' started to compete with earlier terms like 'global change research' or 'earth system science'. The trading zone is clearly visible. While no specific term has become dominant, there are dynamics that affect the configuration of research practices in relation to a wider field of societal concerns.[22]

The World Commission on Environment and Development (WCED 1987) had presented the term 'sustainability' to mark an integrated view on issues of environment and development and the need for coordinated policy strategies. Sustainable development is 'development that meets the needs of the present without compromising the ability of future generations to meet their own', and so required consideration of socio-economic as well as ecological dynamics. Inscribed into this view were the global nature of the challenge and a promise of 'sustainable growth' as a solution to serve both the environment and the economy. As such, the term proved successful in the policy world. In 1992 it was endorsed as an overarching challenge and guiding principle of global public policy at the first 'Earth Summit' in Rio de Janeiro. By the end of the 1990s sustainability had become a global buzzword, and an occasion to consider translation to concrete action.[23]

The surge of 'sustainable development' in the policy discourse also mobilized researchers and science entrepreneurs. As a holistic challenge it called for new approaches of knowledge production. Sustainable development became translated into an epistemic challenge of studying interlinked dynamics of social and ecological systems and how they were to be governed. Scientists started various initiatives to fill the newly opened space with dedicated programmes that went past the established disciplines and their sponsoring arrangements. The International Human Dimensions Programme (IHDP) was set up in 1996 with a view to strengthen the social sciences compared to WCRP and IGBP, two programmes of global change research that had been running already before sustainable development was introduced.[24] The 'Resilience Alliance' built a network of international scientists geared towards the study of what they referred to as social-ecological systems.[25] Such initiatives positioned groups of researchers, and their specific approaches, as knowledge providers for sustainable development. Also universities produced joint declarations which presented themselves as incubators of research for sustainable development and as hosts of education and training programmes.[26] The organizing and positioning of research capacity was undergirded by an abundance of programmatic publications which sought to set out the epistemic agenda of sustainable development (e.g. Norgaard 1994; Schellnhuber and Wenzel 1998; Costanza et al. 1999; Clark et al. 2001; Gunderson and Holling 2002).

Two developments stand out: the declaration of a new 'sustainability science' in 2001 (Kates et al. 2001) and the formation of the 'Earth System Science Partnership' by the global change research programmes.[27] Sustainability science made the stronger epistemic claim, and sought to enrol research practices developed throughout the 1990s, to make a case for fundamentally new concepts and methodologies:

> A new field of sustainability science is emerging that seeks to understand the fundamental character of interactions between nature and society. . . . Combining different ways of knowing and learning will permit different social actors to work in concert, even with much uncertainty and limited information. . . . [It] differs to a considerable degree in structure, methods, and content from science as we know it. . . . In each phase of sustainability science research, novel schemes and techniques have to be used, extended, or invented. . . . Progress in sustainability science will require fostering problem-driven, interdisciplinary research; building capacity for this research; creating coherent systems of research planning, operational monitoring, assessment, and application; and providing reliable, long-term financial support.
>
> (Kates et al. 2001)

The term embodied a promise to develop and maintain links and interactions with the wider world, presenting itself as a bridge between the worlds of knowledge and action:

> [Sustainability Science is] neither 'basic' nor 'applied' research but as a field defined by the problems it addresses rather than by the disciplines it employs; it serves the need for advancing both knowledge and action by creating a dynamic bridge between the two.
>
> (Clark 2007)

As a new candidate umbrella term, competing with 'global change research' or 'earth system science', sustainability science was launched by an international network of scholars,[28] which organized conferences, elaborated joint programmatic statements, and liaised with science policy and funding agencies, so that the term could achieve some consolidation. A scientific journal was established under this title in 2006.[29] The term was picked up by research ministries and funding agencies in several countries. In 2008 it became the title of a stand-alone section in the Proceedings of the US National Academy of Sciences (Clark 2007). Corporate sponsors also referred to the term in organizing their relations with science.[30]

Independent of the efforts of such scientific entrepreneurs, sustainable development functioned as an increasingly forceful reference in the context of science policy. Sustainability-oriented research was part of an agenda to show that science could be activated in the service of broader societal challenges, not only competitiveness and economic growth. In 2002 the US National Research Council commissioned a study, 'Our common journey: A transition towards sustainability' (Board on Sustainable Development, National Research Council 2002), which contained a promise to achieve sustainable development in two generations, provided sufficient resources would be made available for research (Raven 2002: 957). In various locales around the world, priority programmes were established under the responsibility of research agencies or governments.[31] Special centres were also established, such as the Japan Integrated Research System for Sustainability Science (2005), the Stockholm Resilience Centre (2007), or the Institute for Advanced Sustainability Studies in Potsdam (2009). Such programmes, centres, and platforms provided niches in which sustainability research was nurtured, as parts of broader networks and discourses. This is how research became institutionalized to a certain degree, in a rather fragmented manner, and came to depend on coalitions between certain groups of scientists and entrepreneurial sponsors, which allowed established institutions of research funding and science policy profiling to be locally bypassed against the mainstream of economic-growth oriented R&D. There is a grey zone between such dedicated efforts and relabelling of ongoing research as related to sustainability for the sole purpose of increasing eligibility for funding. Furthermore, the epistemic status of sustainability research was contested, especially with respect to its interdisciplinary character and its orientation towards politically defined problems.[32]

At the policy side, the framing of sustainable development as a global problem implied difficulties for translation into support of research. In contrast with political support for 'nanotechnology' or research on the 'ageing society' the sponsoring of scientific activities by reference to sustainability invokes a global public good, not a national or regional one. It thus implies a problem that requires collective action in the area of national or regional science policy making and research funding. This is recognized, and attempts have been made to set up international agreements of cooperation. An International Group of Funding Agencies for Global Change Research (IGFA) met regularly, since the beginning of the 1990s, to coordinate support for international programmes of Global Change Research.

New efforts for mediating science and policy with a view to global sustainability were made in the run up to another 'Earth Summit' in 2012, again held in Rio de Janeiro. The official objective of 'Rio+20' to 'secure renewed political commitment for sustainable development',[33] provided a reason to push further towards the establishment of an integrated knowledge base. Already in 2006, ICSU had started a joint review of global environmental change programmes with the funders in IGFA.[34] This lead into an Earth System visioning process, now together with the International Social Science Council (ISSC), for constructing the agenda of a disciplinary and regionally integrated science for sustainable development (ICSU 2002, 2005; ISSC 2012).[35] Various funding agencies articulated their demands and established a group of 'high-level representatives', the Belmont Forum, to pursue negotiations with representatives of science.[36] In 2010 the Belmont Forum, together with representatives of ICSU and ISSC, and of UNEP, UNESCO, and the United Nations University, met to negotiate a 10-year joint initiative of science policy to '[p]rovide earth system research for sustainable development'. The initiative was finally launched under the label 'Future Earth – research for global sustainability' at the Rio+20 conference.[37]

What we see is convergence towards an interorganizational field while there is still struggle about the preferred umbrella term. There is deliberate negotiation how scientific supply and societal demand can be clinched, as well as how various candidate umbrella terms could be combined to form a phrase that might function as an umbrella. Whether this was just a matter of tactics, or based dedicated reflection, is not clear.

Conclusion and reflections

We identified a phenomenon in the worlds of science, science policy, and general politics: umbrella terms and their concomitant inter-organizational fields, which mediate between ongoing scientific research and policy requirements for societal relevance. We then presented two cases, nanotechnology and sustainability research, which qualified as established and emerging umbrella terms, respectively, and which allowed us to delve into actual complexities. What did we learn? We can compare and contrast the two cases. We can also step back and reflect on what we saw happening, and what this tells us about the dialectics of promising science and technology as modulated by umbrella terms. This will set the scene for a brief discussion of *de facto* governance of science through umbrella terms, and the role of STS scholars in such *de facto* governance.

There are two important differences between the two cases. First, nanotechnology offers open-ended promises for what it might enable us to do, while sustainability science and global change research and earth system science reason back from global challenges to what scientific research should contribute. While the histories are different, the process is the same, with the two cases being at different phases: there are struggles linked to potential umbrella terms, a dominant term emerges and becomes established, at least for some time, as a conduit which allows protection of ongoing research as well orientation towards relevance to societal problems and challenges.

One can zoom in and see an interesting parallel between the group of scientists pushing 'sustainability science' and the group around Drexler pushing nanotechnology as molecular manufacturing. Both have visions about what a 'new kind of science' can achieve, and both get a hearing. In the case of nanotechnology, the clinching of supply and demand came from another direction, with the US National Nanotechnology Initiative and its international repercussions, which overtook (and eclipsed) the Drexlerian vision. In the case of sustainability science, the ambitions may also be too high, but the sustainability scientists (to coin a term in much the same way that the term nanoscientists emerged) appear to be well embedded in international establishment organizations and networks. They may make some progress in the coming years, even if more technocratic versions have to be accommodated in ongoing negotiations with disciplinary scientists and policy makers, as is visible in complementary references to 'Earth System Science'.

A hard-nosed question, for both cases, is whether umbrella terms just reflect the latest fashion in science funding and sponsorship, and will be washed away when the next wave arrives. The umbrella term may disappear, but there will be lasting structural changes, linked to inter-organizational fields that emerged and solidified. In the meantime, actors in the worlds of science and of science policy will use actual and potential umbrella terms for their own purposes. But once an umbrella term is in place, i.e. after the clinching of supply and demand and some institutionalization, it cannot be escaped (or only at a cost). So in addition to their indicating a new pattern of science governance which combines relevance considerations and some autonomy of research (as befits the regime of Strategic Science), the term itself has a governance effect. Umbrella terms, once established, are a de facto governance technology, and actors realize this and struggle about the term and its articulation.[38] The eventual result, an umbrella term becoming forceful, is the outcome, at a collective level, of many actions and interactions.

Thus, there are two ways that umbrella terms are a governance technology: they constitute an arena for struggles, about definitions and access/exclusion, about resources;[39] and their eventual black-boxed use has effects exactly because the detailed struggles that went into them are eclipsed.

Two final reflections are in order. First, about governance of science. While use of the term governance helps to move away from an exclusive focus on government and attempts at top-down steering, there is still a top-down bias in many studies, in the sense that government steering is the standard, which is now to be modified. What we have shown is that there are elements of governance of science in ongoing developments, exemplified in this paper by the emergence and stabilization of umbrella terms mediating between science, science policy, and society. Governance then shifts from attempts to realize policy goals as such, to considering what is happening anyway and how this is modulated in reference to public interests.

Second, about the role of STS scholars. Both authors were and are active in the fields we used as case studies in this article, up to benefiting from the new resource flows by having their own research projects funded. We had discussions with actors in the field, and sometimes explicitly (even if modestly) intervened.[40] The present article is a further move: it opened up the black box of umbrella term dynamics – a typical STS approach – and if it were to be read by actors in the field, they could take it up as a move in their struggles. But we also contribute to the existence of the field, because talking about 'nanotechnology' or 'sustainability science' helps make them become more real. This is unavoidable, and one should not retract from it,[41] but try to understand what is happening and position oneself reflexively.

Notes

1 There have been studies of governance of science by STS scholars all along, but they were considered to be at the margin of the field. This is changing now; see for example the shift in contents of the two STS Handbooks (Jasanoff et al. 1995; Hackett et al. 2008). In 1995, all the classical themes of STS research were present, and one of the seven parts of the Handbook discussed science, technology, and the State, with an emphasis on trends to be observed rather than governance questions. In 2008, two of the five parts were devoted to such issues, often explicitly discussing governance. See also the collection of papers in Jasanoff (2004).

2 The same comment can be made about STS scholars getting involved in the recent wave of technology assessment and ELSA studies of nanotechnology, and is being made, as one of us (AR) can testfiy.

3 So-called strategic research programmes started to be drawn up and implemented already in the 1970s (Rip 1990; Rip and Hagendijk 1988),

4 See Galison's (1997) discussion of 'trading zones'. He considered mutual translations between different disciplines and fields of research which would lead to emergence of pidgins and creoles. In our discussion, the translations are between fields of science and science policy, and society as a further reference. The point about emergence of pidgins and creoles remains applicable, up to the emergence of a 'blizzard of buzzwords' (Ziman 1994) that is part of the regime of Strategic Science. The recontextualization of science in society is genuine, however (Nowotny et al. 2001; Markus et al. 2009).

5 In the example of nanotechnology, the fact that it covers a wide variety of scientific approaches and technological options is recognized. After noting that nanotechnology 'has become a handy shorthand label for several phenomena', Hodge et al. (2010: 6) discuss 'the immense range of technologies that fall under the nanotechnologies umbrella'. Indicative also is how the European Commission and the UK Research Councils now speak of nanotechnologies in the plural.

6 As societal concerns for relevance are sought to be embodied in the organization of the field, specific conceptions of society and its problems that are underlying the notion of 'challenges' become inscribed into the emerging configuration of social relations under the umbrella. As it becomes an institutional reality, an umbrella term may thus 'co-produce' a particular form of science with a particular politically articulated form of society. On this point see, for example, Miller's (2004) analysis of interrelations between the constitution of a science of the global climate with the constitution of a new global political order.

7 Here, we use the term 'mediator' in a commonsensical way, but we can also refer to Latour and to Callon. In Actor-Network Theory, mediators are circulating entities with an inside that can be 'read' in and through their action. Callon (1991), who speaks of intermediaries in the sense of what Latour (2005) called mediators, gave examples of texts (inscriptions), technical artefacts, human bodies, and money (and other promissories).

8 What we describe here is a central dynamic of priority setting, where supply and demand meet and become entangled in their further articulation in a variety of ways.

9 In the trading zone between 'relevance' and 'ongoing science', authority to translate, in the process of emergence of umbrella terms and in their eventual institutionalization, will thus allow exerting power, for resource mobilization and for research governance. Struggles about definition and scope of the field, which are very visible in nanotechnology, are struggles to become authoritative.

10 To be sure, the notion of a 'clinching event' is retrospective: whether an event is 'clinching' will not be clear at the time. Depending on further developments in the area of sustainability and science, one or another present event may turn out to have 'clinched' supply and demand.

11 The term 'nanotechnology' was coined by Taniguchi (1974), for his own purposes. He is duly referenced, but his definition is not taken up.

12 Neither the press release (http://nobelprize.org/nobel_prizes/physics/laureates/1986/press.html) nor the acceptance speech by Binnig and Rohrer mention nanotechnology or nanoscale science. They locate their work with respect to surface science. (They do mention, at the very end, that their scanning tunnelling microscope might be used to move atoms, and thus work as a 'Feynman machine'; Binnig and Rohrer 1993: 407). Ten years later, the new laureates (as well as the press release) still focus on the science, now of fullerenes, but do make a reference to what happens 'at the nanotechnology front' (Kroto 2003: 76).

13 For example, the very early UK National Initiative on Nanotechnology was an awareness-raising initiative, primarily in terms of market potential of the new research results, but could not generate industrial interest. Apart from two small activities, it was quiet on the nano front in the UK until the end of the 1990s. (Van der Most 2009: 59). In 1996, the UK Parliamentary Office of Science and Technology published an overview of possible applications, under the title *Making it in Miniature: Nanotechnology, UK Science and its applications*, but was content to note improvements in miniaturization of chips, in sensors, in surfaces, in diagnostic tools (ibid.: 6).

14 Running ahead of the story: when the label nanotechnology became institutionalized (almost overnight, with the announcement of the US NNI), it became important to define its scope and who could legitimately refer to the label. Thus, Drexler's futuristic project had to be excluded from what was now to be the mainstream. It became common to refer to molecular manufacturing as 'science fiction'. The 2003 (orchestrated) debate between Drexler and Smalley on the feasibility of molecular manufacturing has become iconic. Drexler countered the mainstream moves by calling this work superficial rather than deep nanotechnology, and so claimed 'real' nanotechnology for himself. He lost the struggle, though (Rip and van Amerom 2010).

15 Also dedicated journals like Journal of Nanoparticle Research (since 1999), Journal of Micro-Nano Mechatronics (since 2004). Grieneisen (2010) notes the exponential growth, since the end of 1990s, and definitely since 2005, of journals devoted to nanotechnology. The first journal devoted exclusively to nanoscale science and technology, *Nanotechnology*, was launched by Institute of Physics Publishing in July 1990. During the 1990s, only a few 'nano-journals' were launched; by 1998, the total number was 18. By 2010, 165 'nano-journals' had been launched, and 142 were still producing.

16 He actually called for the use of nanotechnology as an umbrella term: 'Nanotechnology, Smalley concluded, presented a "tremendously promising new future." What was needed was someone bold enough to put a flag in the ground and say: "Nanotechnology, this is where we are going to go"'. (McCray 2005: 187).

17 Innovation did occur, in micro-nano-electronics and with nanomaterials and nanostructured surfaces for mundane but useful applications like coatings, dirt-repellent textiles, and reinforced tyres and tennis rackets.

18 In the Netherlands, the NanoNed R&D Consortium (2003–2010), funded by public money, framed its research themes as basic research with perhaps some applications. Its successor, NanoNextNL, again funded by public money and some industrial contributions, had to frame a large part of its research in relation to energy, water, health, and food. There was also political pressure to have 15 percent of the budget spent on research directly or indirectly related to possible risks of nanotechnology. For Cannoned, see www.nanoned.nl/. For its successor, NanoNextNL, see www.nanonextnl.nl/

19 The notion of 'technoscientific objects' is the topic of a recent research project led by Alfred Nordmann and Bernadette Bensaude-Vincent. Available at: www.philosophie. tu-darmstadt.de/goto/goto/home/home.en.jsp.

20 There is a history of the rise of terms like 'the environment' and 'environmental' in the 1970s, which functioned to some degree as an umbrella term under which funding programmes and university degrees were taking shape. Such use of the term 'the environment' continues, as in the title of Lubchenco's (1998) article: 'Entering the century of the environment'. Scientific unions, rooted mainly in the natural sciences, played a crucial role in articulating 'the environment' and its threat of deterioration or collapse. Prominent efforts were the 1972 Report to the Club of Rome on 'the limits to growth' (Meadows et al. 1972) in connection with the first UN conference on the Environment in Stockholm in 1972, and its repercussions (e.g. establishment of UN Environment Programme and Environmental Ministries in many nation states).

21 The notion of an ideograph was introduced by McGee (1980) to capture the function of terms like 'the people', that are diffusely defined, allow various meanings to be projected onto them and are important to capture in a debate because of their positive

rhetorical value. Rip (1997) showed how 'industry' and 'sustainability' functioned as ideographs in science policy discussions and practices. The same hold for 'sustainable development', and is not limited to science policy occurrences.

22 When using the term 'sustainability research' as the heading of this section of the paper, we might be seen as taking sides in the struggle. Since we needed a simple heading, we chose the one which is relatively neutral compared with the other possibilities.

23 There is an ongoing battle over precise definitions and concrete actions which reflect a continued struggle for dominance between ecological and economic concerns, North and South, global and local – all those oppositions which 'sustainable development', as a political term, sought to overcome (Voß and Kemp 2006).

24 In 1979 the World Climate Research Programme (WCRP) was established (with sponsorship by the World Meteorological Organisation, WMO, and the International Council of Scientific Unions, ICSU), leading up to the 'Toronto Conference on the Changing Atmosphere' in 1988 (paving the way for the Intergovernmental Panel on Climate Change, IPCC, and subsequent negotiations of a UN Convention on Climate Change). A broader focus on the global environment, and how it changed, was adopted by the International Geosphere-Biosphere Programme (IGBP) which was established in 1986, also sponsored by ICSU.

25 The Alliance was established in 1999, see www.resalliance.org

26 For example the 1990 Talloires Declaration of University Presidents for a Sustainable Future; the 1993 Kyoto Declaration on Sustainable Development by the International Association of Universities (IAU). This continued, see for one example the July 2008, G8 University summit ('27 of the leading educational and research institutions in the G8 member nations') producing the 'Sapporo Sustainability Declaration' (available at: http://g8u-summit.jp/english/ssd/); Alliance for Global Sustainability (available at: http://globalsustainability.org/)

27 In 2001, the international research programmes on global environmental change (WCRP, IGBP, IHDP, plus a newly established one on biodiversity, Diversitas) got together under the umbrella of the Earth System Science Partnership (ESSP). Their 'Amsterdam Declaration' stated, 'The business-as-usual way of dealing with the Earth System is not an option. It has to be replaced ¬ as soon as possible ¬ by deliberate strategies of good management that sustain the Earth's environment while meeting social and economic development objectives. . . . A new system of global environmental science is required. This is beginning to evolve from complementary approaches of the international global change research programmes and needs strengthening and further development. It will draw strongly on the existing and expanding disciplinary base of global change science; integrate across disciplines, environment and development issues and the natural and social sciences; collaborate across national boundaries on the basis of shared and secure infrastructure; intensify efforts to enable the full involvement of developing country scientists; and employ the complementary strengths of nations and regions to build an efficient international system of global environmental science'.

28 Its stronghold is at the Program of Sustainability Science at Harvard University's Center for International Development. See: www.hks.harvard.edu/centers/cid/programs/sustsci (see also Board on Sustainable Development 2002).

29 *Sustainability Science*, established under the auspices of Springer Japan, introduces itself by the editorial notes: 'Sustainability Science provides a trans-disciplinary platform for contributing to building sustainability science as a new academic discipline focusing on topics not addressed by conventional disciplines. As a problem-driven discipline, sustainability science is concerned with practical challenges such as those caused by climate change, habitat and biodiversity loss, and poverty. At the same time it investigates root causes of problems by uncovering new knowledge or combining current knowledge from more than one discipline in a holistic way to enhance understanding of sustainability'.

30 cf. the 2010 International Conference on Sustainability Science (sponsored by business corporations and set up with a view to further links between 'world scientific leaders in Sustainability Science and representatives from industry and civil society', see http://icss2010.net/?p=industry-profiles), or the journal SAPIENS, which is sponsored by the transnational company Veolia to publish review articles and evidence-based opinions that integrate knowledge across disciplines.

31 At the European Union, DG Research (now DG Research and Innovation) hosts a platform for 'sustainability science' and launched an initiative Research and Development for Sustainable Development (RD4SD) including a Conference on 'Sustainable development: a challenge for European research' in 2009. The German Research Foundation (DFG) had a 'Schwerpunktprogramm Mensch und globale Umweltveränderung' (www4.psycholo gie.uni-freiburg.de/umwelt-spp/welcome.html), the German Federal Ministry for Education and Research (BMBF) set up a funding initiative for 'social-ecological research' (www.sozial-oekologische-forschung.org/) in 2000, and later established 'research for sustainable development' (Fona) as an umbrella label for a variety of research lines that were drawn together onto a common 'platform' (www.fona.de/).

32 There is a tension between natural and social sciences, cf. 'Sustainability science has a good deal to say about how we can logically approach the challenges that await us, but the social dimensions of our relationships are also of fundamental importance' (Leshner 2002: 957). There are also discussions about methods and quality criteria of sustainability science as a normatively oriented endeavour aspiring to inclusiveness with regard to a diversity of knowledges that are to be integrated (e.g. Thompson Klein et al. 2001; Nölting et al. 2004; Bergmann et al. 2005; Pohl and Hirsch-Hadorn 2007).

33 www.uncsd2012.org/objectiveandthemes.html

34 It was concluded, 'There is a clear need for an internationally coordinated and holistic approach to Earth system science that integrates natural and social sciences from regional to the global scale' (ICSU-IGFA, 2008), further that there is a 'need for a unified strategic framework . . . to deepen understanding . . . deliver solutions'.

35 ICSU co-sponsored all programmes of global environmental change research as well as coordinated efforts on 'joint projects on global sustainability' (in Water, Food, Carbon, Human Health) under the Earth System Science Partnership. With promoting IHDP, since 1996, the Council undertook targeted efforts to give a role to the social sciences (see ISSC 2012). The Earth System visioning (2009–2011) articulated research questions as 'five grand challenges' from the point of view of science: 'observing, forecasting, thresholds, responding, innovating'.

36 The Belmont Forum, established in 2009 out of IGFA: 'A high level group of the world's major and emerging funders of global environmental change research and international science councils [which] acts as a Council of Principals for the broader network of global change research funding agencies, IGFA [so] aligning international resources' indicates a further attempt at creating an interorganizational field. '[It] developed a collective "funders" vision of the priorities for global environmental change research' (Belmont Forum 2011). Cognitive challenges are identified, linked with action perspectives – and a candidate umbrella term: 'recognition that the understanding of the environment and human society as an interconnected system, provided by Earth System research in recent decades . . . to provide knowledge for action and adaptation to environmental change . . . remove critical barriers to sustainability . . . integrated into a seamless, global Earth System Analysis and Prediction System (ESAPS), which will provide decision-makers with a holistic decision support framework' (ibid.).

37 The declared aim of 'Future Earth' is 'reorganizing the entire global environmental change research structure, and the way of doing research' in a view of 'integrating the understanding of how the Earth system works to finding solutions for a transition to global sustainability'. It seeks to build on and integrate earlier activities 'and enhance . . . global environmental change programmes and projects', but with a view towards 'new solution focused projects'. The approach is one of 'co-designing and

co-producing research agendas and knowledge' by 'policy makers, funders, academics, business and industry, and other sectors of civil society' (ICSU 2012).

38 This is part of a larger problem, which one of us (JPV) has articulated for the case of policy instruments as a governance technology: at the knowledge production side there is linking and packaging to create in input in policy (as provision of solutions), which then somehow functions in the making and implementation of policy (as treatment of public problems) (Voß 2007b, 2007a).

39 A similar point is made by Wullweber (2008) for nanotechnology, using Laclau's notion of an 'empty signifiers' (Laclau 1996).

40 As we did in our projects of constructive technology assessment of nanotechnology, we have conceptualized this as 'insertion', see Rip and Robinson (forthcoming).

41 This is an argument against Latour's position: 'The task of defining and ordering the social should be left to the actors themselves, not taken up by the analyst'. 'ANT simply doesn't take as its job to stabilize the social on behalf of the people it studies; such a duty is to be left entirely to the "actors themselves"' Latour (2005: 23, 30).

References

Bartels, Ditta, 1984: The Rockefeller Foundation's Funding Policy for Molecular Biology: Success or Failure? *Social Studies of Science*, Vol. 14, 238–243.

Belmont Forum, 2011: *The Belmont Challenge: A Global, Environmental Research Mission for Sustainability*. White paper.

Bergmann, Matthias, et al., 2005: *Quality Criteria for Transdisciplinary Research: A Guide for the Formative Evaluation of Research Projects*. Frankfurt am Main: ISOE.

Binnig, Gerd and Heinrich Rohrer, 1993: Scanning Tunneling Microscopy: From Birth to Adolescence: Nobel Lecture, December 8, 1986. In Gösta Ekspång (ed.) and Tore Frängsmyr (ed. in Charge), *Nobel Lectures, Physics 1981–1990*. Singapore: World Scientific Publishing, 389–409.

Board on Sustainable Development, National Research Council, 2002: *Our Common Journey: A Transition toward Sustainability*. Washington, DC: National Academy Press.

Böhme, Gernot, et al., 1978: *Die gesellschaftliche Orientierung des wissenschaftlichen Fortschritts*. Frankfurt am Main: Suhrkamp.

Bonaccorsi, Andrea, 2008: Search Regimes and the Industrial Dynamics of Science. *Minerva*, Vol. 46, 285–315.

Bonaccorsi, Andrea, 2010: *Institutional Governance of Science in Fast Moving Scientific Fields*. Paper presented to the Conference on Changing Governance of Research, Frankfurt, March 11–12.

Bush, Vannevar, 1945: *Science: The Endless Frontier: A Report to the President on a Program for Postwar Scientific Research*. Reprinted, with Appendices and a Foreword by Daniel J. Kevles, by the National Science Foundation, Washington, DC, 1990.

Callon, Michel, 1991: *Techno-Economic Networks and Irreversibility*. In John Law (ed.), *A Sociology of Monsters: Essays on Power, Technology and Domination*. London: Routledge, 132–161.

Clark, William C., Jill Jäger and Josee van Eijndhoven, 2001: *Managing Global Environmental Change: An Introduction to the Volume. Learning to Manage Global Environmental Risks. G. The Social Learning*. Cambridge, MA: MIT Press, 1–20.

Clark, William C., 2007: Sustainability Science: A Room of Its Own. *Proceedings of the National Academy of Science*, Vol. 104, 1737–1738.

Costanza, Robert, et al. (eds.), 1999: *Institutions, Ecosystems and Sustainability*. Cambridge: Cambridge University Press.

Dolby, Robert G.A., 1976: The Case of Physical Chemistry. In Gérard Lemaine, et al. (eds.), *Perspectives on the Emergence of Scientific Disciplines*. The Hague: Mouton/ Aldine, 63–73.

Drexler, K. Eric, 1986: *Engines of Creation: The Coming Era of Nanotechnology*. New York: Anchor Press/Doubleday.

Ede, Andrew, 2007: *The Rise and Decline of Colloid Science in North America, 1900– 1935: The Neglected Dimension*. Aldershot: Ashgate.

Etzkowitz, Henry and Loet Leydesdorff, 2000: The Dynamics of Innovation: From National Systems and 'Mode 2' to a Triple Helix of University-Industry-Government Relations. *Research Policy*, Vol. 29, 109–123.

Funtowicz, Silvio and Jerome Ravetz, 1993: Science for the Post-Normal Age. *Futures*, Vol. 25, 735–755.

Galison, Peter, 1997: *Image and Logic: A Material Culture of Microphysics*. Chicago: University of Chicago Press.

Gibbons, Michael, et al., 1994: *The New Production of Knowledge: The Dynamics of Science and Research in Contemporary Societies*. London: Sage.

Grieneisen, Michael L., 2010: The Proliferation of Nano Journals. *Nature Nanotechnology*, Vol. 5, 825.

Gunderson, Lance H. and Crawford S. Holling, 2002: *Panarchy: Understanding Transformations in Human and Natural Systems*. Washington, DC: Island Press.

Hackett, Edward J., Olga Amsterdamska, Michael Lynch and Judy Wacjman (eds.), 2008: *The Handbook of Science and Technology Studies*. Cambridge, MA: MIT Press.

Hodge, Graeme A., Diana M. Bowman and Andrew Maynard, 2010: Introduction: The Regulatory Challenges for Nanotechnologies. In: G.A. Hodge, D.M. Bowman and A. Maynard (eds.), *International Handbook on Regulating Nanotechnologies*. Cheltenham: Edward Elgar, 1–24.

ICSU (International Council for Science), 2002: *Science and Technology for Sustainable Development*. Consensus Report and Background Document, Mexico City Synthesis Conference, May 20–23, 2002, ICSU Series on Science for Sustainable Development No. 9, Paris, 30 pp. www.icsu.org/Gestion/img/ICSU_DOC_DOWNLOAD/70_DD_FILE_Vol9.pdf

ICSU (International Council for Science), 2005: *Harnessing Science, Technology, and Innovation for Sustainable Development*. Report from the ICSU-ISTS-TWAS Consortium Ad Hoc Advisory Council, 38 pp. www.icsu.org/Gestion/img/ICSU_DOC_DOWNLOAD/584_DD_FILE_Consortium_Report.pdf

ICSU (International Council for Science), 2012: *Future Earth: Research for Global Sustainability*. Leaflet. www.icsu.org/future-earth/whats-new/relevant_publications/future-earth-leaflet

ICSU-IGFA (International Council for Science-International Group of Funding Agencies for Global Change Research), 2008: *Review of the Earth System Science Partnership (ESSP)*. www.essp.org/fileadmin/redakteure/pdf/others/ESSP_Review_08.pdf

Irvine, John and Ben R. Martin, 1984: *Foresight in Science: Picking the Winners*. London: Frances Pinter.

ISSC (International Social Science Council), 2012: *Transformative Cornerstones of Social Science Research for Global Change*. International Social Science Council, 3 May.

Jasanoff, Sheila, Gerald E. Markle, James C. Petersen and Trevor Pinch (eds.), 1995: *Handbook of Science and Technology Studies*. Thousand Oaks, CA: Sage Publications.

Jasanoff, Sheila, 2004: *States of Knowledge: The Co-Production of Science and Social Order*. London: Routledge.

Kates, Robert W., et al., 2001: Sustainability Science. *Science*, Vol. 292, 641–642.

Kohler, Robert E., 1976: The Management of Science: The Experience of Warren Weaver and the Rockefeller Foundation Programme. *Minerva Biotecnologica*, Vol. 14, 279–306.

Konrad, Kornelia, 2004: *Prägende Erwartungen. Szenarien als Schrittmacher der Technikentwicklung*. Berlin: edition sigma.

Konrad, Kornelia, 2006: The Social Dynamics of Expectations: The Interaction of Collective and Actor Specific Expectations on Electronic Commerce and Interactive Television. *Technology Analysis & Strategic Management*, Vol. 18, 429–444.

Kroto, Harold W., 2003: Symmetry, Space, Stars and C60: Nobel Lecture, December 7, 1996. In Ingmar Grenthe (ed.), *Nobel Lectures, Chemistry 1996–2000*. Singapore: World Scientific Publishing, 44–79.

Kuhn, Thomas S., 1970: *The Structure of Scientific Revolutions*, second enlarged edition. Chicago: University of Chicago press.

Laclau, Ernesto, 1996: Why Do Empty Signifiers Matter to Politics? In Ernesto Laclau (ed.), *Emancipation(s)*. London: Verso, 34–46.

Latour, Bruno, 2005: *Reassembling the Social: An Introduction to Actor-Network Theory*. New York: Oxford University Press.

Lave, Rebecca, Philip Mirowski and Samuel Randalls, 2010: Introduction: STS and Neoliberal Science. *Social Studies of Science*, Vol. 40, 659–675.

Leshner, Alan, 2002: Science and Sustainability. *Science*, Vol. 297, 897.

Lubchenco, Jane, 1998: Entering the Century of the Environment: A New Social Contract for Science. *Science*, Vol. 279, 491–497.

Markus, Eszter, et al., 2009: *Challenging Futures of Science in Society: Emerging Trends and Cutting-Edge Issues*. Report of the MASIS Expert Group to the European Commission, October.

McCray, W. Patrick, 2005: Will Small Be Beautiful? Making Policies for Our Nanotech Future. *History and Technology*, Vol. 21, 177–203.

McGee, Michael C., 1980: 'The Ideograph': A Link Between Rhetoric and Ideology. *The Quarterly Journal of Speech*, Vol. 66, 1–16.

Meadows, Donella H., et al., 1972: *The Limits to Growth: A Report to the Club of Rome*. New York: Universe Books.

Miller, Clark A., 2004: Climate Science and the Making of a Global Political Order. In S. Jasanoff (ed.), *States of Knowledge: The Coproduction of Science and Social Order*. London: Routledge, 66–76.

Miller, Richard, 2009: From Ageing Research to Preventive Medicine: Pathways and Obstacles. In P. Healey and S. Rayner (eds.), *Unnatural Selection: The Challenges of Engineering Tomorrow's People*. London: Earthscan.

Nölting, Benjamin, Jan-Peter Voß and Doris Hayn, 2004: Nachhaltigkeitsforschung – jenseits von Diziplinierung und 'anything goes'. *Gaia*, Vol. 13(4), 254–261.

Norgaard, Richard B., 1994: *Development Betrayed: The End of Progress and a Coevolutionary Revisioning of the Future*. London: Routledge.

Nowotny, Helga, Peter Scott and Michael Gibbons, 2001: *Re-Thinking Science: Knowledge and the Public in an Age of Uncertainty*. Cambridge: Polity Press.

OECD (Organisation for Economic Co-Operation and Development), 1971: *Science, Growth and Society: A New Perspective*. Paris: OECD.

Pohl, Christian and Gertrude Hirsch-Hadorn, 2007: *Principles for Designing Transdisciplinary Research*. Munich: Oekom.

Raven, Peter H., 2002: Science, Sustainability, and the Human Prospect. *Science*, Vol. 297, 954–958.

RCUK (Research Council UK), 2009: *RCUK Delivery Plan 2008/09 to 2010/11*. www.rcuk. ac.uk/documents/publications/anndeliveryplanrep2008-09.pdf

Regis, Ed, 1996: *Nano: The Emerging Science of Nanotechnology*. New York: Back Bay Books.

Rettig, Richard A., 1977: *Cancer Crusade: The Story of the National Cancer Act of 1971*. Princeton, NJ: Princeton University Press.

Rip, Arie, 1990: An Exercise in Foresight: The Research System in Transition: To What? In S.E. Cozzens, P. Healey and J. Ziman (eds.), *The Research System in Transition*. Dordrecht: Kluwer Academic, 387–401.

Rip, Arie, 1997: A Cognitive Approach to Relevance of Science. *Social Science Information*, Vol. 36, 615–640.

Rip, Arie, 2002: Regional Innovation Systems and the Advent of Strategic Science. *Journal of Technology Transfer*, Vol. 27, 123–131.

Rip, Arie, 2006: A Co-Evolutionary Approach to Reflexive Governance: And Its Ironies. In J.-P. Voß, D. Bauknecht and R. Kemp (eds.), *Reflexive Governance for Sustainable Development*. Cheltenham, UK: Edward Elgar, 82–100.

Rip, Arie, 2011: Science Institutions and Grand Challenges of Society: A Scenario. *Asian Research Policy*, Vol. 2, 1–9.

Rip, Arie, 2012: The Context of Innovation Journeys. *Creativity and Innovation Management*, 21(2), 158–170.

Rip, Arie and Rob Hagendijk, 1988: *Implementation of Science Policy Priorities*. SPSG Concept Paper No. 2. London: Science Policy Support Group, March.

Rip, Arie and Douglas K.R. Robinson, forthcoming: Constructive Technology Assessment and the Methodology of Insertion. In Neelke Doorn, et al. (eds.), *Early Engagement and New Technologies: Opening Up the Laboratory*. Dordrecht: Springer Science Business & Media, 2013.

Rip, Arie and Marloes van Amerom, 2010: Emerging de facto Agendas Around Nanotechnology: Two Cases Full of Contingencies, Lock-Outs, and Lock-Ins. In Mario Kaiser, et al. (eds.), *Governing Future Technologies: Nanotechnology and the Rise of an Assessment Regime*. Dordrecht: Springer, 131–155.

Schäfer, Wolf (ed.), 1983: *Finalization in Science: The Social Orientation of Scientific Progress*. Dordecht: D. Reidel.

Schellnhuber, Hans-Joachim and V. Wenzel, 1998: *Earth System Analysis: Integrating Science for Sustainability*. Complemented Results of a Symposium Organized by the Potsdam Institute (PIK). Berlin: Springer.

Taniguchi, N., 1974: *On the Basic Concept of Nanotechnology*. Proceedings of the International Congress on Production Engineering, JSPE, Tokyo.

Thompson Klein, Julie, et al., 2001: *Transdisciplinarity: Joint Problem Solving Among Science, Technology, and Society: An Effective Way for Managing Complexity*. Berlin: Birkhäuser.

Van den Daele, Wolfgang, Wolfgang Krohn and Peter Weingart, 1979: *Geplante Forschung. Vergleichende Studien über den Einfluss politischer Programme auf die Wissenschaftsentwicklung*. Frankfurt/Main: Suhrkamp.

Van der Most, Frank, 2009: *Research Councils Facing New Science and Technology: The Case of Nanotechnology in Finland, the Netherlands, Norway, and Switzerland*. PhD Thesis. Enschede: University of Twente.

Van Lente, Harro and Arie Rip, 1998: Expectations in Technological Developments: An Example of Prospective Structures to be Filled in by Agency. In Cornelis Disco and

Barend van der Meulen (eds.), *Getting New Things Together*. Berlin and New York: Walter de Gruyter, 195–220.

Voß, Jan-Peter, 2007a: *Designs on Governance: Development of Policy Instruments and Dynamics in Governance*. PhD Thesis. Enschede: University of Twente, School of Management and Governance.

Voß, Jan-Peter, 2007b: Innovation Processes in Governance: The Development of 'Emissions Trading' as a New Policy Instrument. *Science and Public Policy*, Vol. 34(5), 329–343.

Voß, Jan-Peter and René Kemp, 2006: Sustainability and Reflexive Governance: Introduction. In J.-P. Voß, D. Bauknecht and R. Kemp (eds.), *Reflexive Governance for Sustainable Developement*. Cheltenham, UK: Edward Elgar, 3–28.

WCED (World Commission on Environment and Development), et al., 1987: *Our Common Future*. New York: United Nations.

Wullweber, Joscha, 2008: Nanotechnology: An Empty Signifier à venir? A Delineation of a Techno-Socio-Economical Innovation Strategy. *Science, Technology & Innovation Studies*, Vol. 4, 27–45.

Ziman, John, 1994: *Prometheus Bound: Science in a Dynamic Steady State*. Cambridge: Cambridge University Press.

3   Dual dynamics of promises and waiting games around emerging nanotechnologies

Alireza Parandian, Arie Rip,
and Haico te Kulve

Published as

Dual dynamics of promises and waiting games around emerging nanotechnologies, *Technology Analysis & Strategic Management*, 24(6) July 2012, 565–582

Introduction

Newly emerging sciences and technologies (NEST) are accompanied by promises about their interest, performance, and societal effects. At the same time, there are uncertainties about eventual performance and embedding in society. Addressing such uncertainties in strategic management and policy requires estimates of future developments, but also insight in the present situation and the forces shaping the ongoing dynamics, ranging from industry structures and governance regimes to less tangible but still forceful cultural factors.

These observations are uncontroversial. This paper adds to them by drawing attention to a paradox. New technologies like nanotechnology are surrounded by big promises, envisioning a third industrial revolution, environmental remediation, and human enhancement. This 'halo' mobilizes policy actors, industrialists, and publics.[1] At the same time, the existence of these promises can be an obstacle to the realization of these envisaged benefits because of their diffuse and open-ended nature: this makes actors uncertain about directions to go and thus creates reluctance to invest in concrete developments. To highlight that this is a structural problem, we will speak of 'waiting games'.

When tracing developments in two fields with strong nanotechnology promises, organic large area electronics (OLAE) and nanotechnology-enabled targeted drug delivery, we encountered a striking phenomenon: the continuing reluctance of many actors to invest money and effort in developments, despite recognizing the promises. Thus, promises are not turned into requirements that guide concrete development trajectories (cf. Van Lente 1993; Van Lente and Rip 1998, on promise-requirement cycles).

Clearly, there are two different promise dynamics: those of big but open-ended promises, which we call 'umbrella promises', and the more concrete promise-requirement cycles. They are not independent, and one can speak of dual dynamics of promises.[2]

The first part of the paper will develop this observation conceptually. The pattern (or de facto arrangement) that we are interested in is the waiting game linked to dual dynamics of promises. We argue that this is a recurrent pattern, over and above the waiting strategies that different actors may follow. In the second part of the paper, we will offer empirical data from our two fields, showing that there are indeed waiting games driven by the force of open-ended promises. In addition, we will briefly discuss how different kinds of actors and actor-collectives address waiting games.

The issue of waiting games linked to dual dynamics of promises is felt and recognized by actors, yet surprisingly has received little scrutiny in the literature.[3] The understanding we offer of these processes and patterns can be used to develop management and governance intelligence.

Conceptualization: dual dynamics of promises and waiting games

When developing our overall conceptualization and considering different types of promise dynamics, we will not pay much attention to concrete actors and contexts. These are important. Researchers, firms, governmental, and non-governmental actors play different roles and will perceive – and voice – expectations differently. Also, actors are tied up in a web of relationships with each other, which enables and constrains their actions and interactions, independent of promises of emerging technologies. Although such broader aspects will be backgrounded to some extent in the discussion of our conceptual framework, they will be visible in our case studies.

For our conceptual framework we can build on the growing literature on sociology of expectations (or expectation statements) in innovation processes and the embedding of new and emerging science and technologies in society (Van Lente 1993; Van Lente and Rip 1998; Brown and Webster 2000; Geels and Smit 2000; Brown and Michael 2003; Konrad 2004; Borup et al. 2006). For newly emerging science and technology (NEST), expectations are functionally important because it is through them that future value of technological options is articulated and to some extent becomes stabilized. In general, expectations motivate action to intervene in socio-technical networks and transform the current 'state of the world' as economists would say.[4] Expectations, when shared, allow some coordination, and there are now attempts at joint coordination of emerging technologies and their future application, as in European Technology Platforms.[5]

Umbrella promises

The general perspective is well established by now. What has got less attention is that there are different patterns of dynamics of promises, and what the differences imply in the real world, including the paradoxical situation of big promises lead-ing to waiting games. Kornelia Konrad's (2004, 2006) work is an exception, and we have been inspired by her analysis of, what we label as, the ' umbrella prom-ise' of the information super highway in the early 1990s, and how it could survive in spite of continual failure of concrete projects.

Umbrella promises are open-ended and can remain very general in nature. A variety of more specific promises can be subsumed under these 'umbrella' prom-ises. They are primarily a discursive phenomenon, and offer narratives linking a new field to broad problems like competition in the world economy, and societal and global issues like climate change. Such narratives fit into the recent policy discourse of 'grand challenges' (Lund Declaration 2009), and legitimize the rel-evance of the promise to the wider world. Occasionally concerns about technolo-gies getting out of control are also voiced in such narratives. When accepted, an umbrella promise (and sometimes black-boxed through the use of an umbrella term like 'nanotechnology') functions as protection of more specific promises and whatever happens on the ground.[6] To maintain their relative freedom, actors may well be reluctant to further specify the open-ended and diffuse promise.

In the particular constellation of the present regime of economics of techno-scientific promise there is a further dynamic: the supply and demand of promises (Joly, Rip and Callon 2010). Narratives and visions are formulated by scientists who need to mobilize resources. Eventual realization of their vision is only one and sometimes only a minor goal. At the side of policy and politics there is a demand because policy actors have to fill a portfolio of interesting initiatives and priorities to gain political support for their role and to justify spending on R&D. The producers of narratives and visions, scientists or other actors, may antici-pate political agendas and strategically formulate umbrella promises. If successful they become part of what Konrad (2006, 431) calls a 'generalized and taken-for-granted social repertoire' (of promises and promising). Joly, Rip and Callon (2010) emphasize how promises are presented so as to create a sense of urgency: referring to the need not to lag behind in competitive environments (scientific, technological, and economic) and/or the need to address 'grand challenges'.[7]

The open-ended umbrella promises become embodied in the form of activities, like funding decisions and their justification, emerging networks, and attempts to improve coordination of technology development and application. These are material indicators of their existence and importance, see Van Lente (1993, 193).

Promise-requirement cycles

Promises can be more specific, i.e. be formulated in terms of a concrete per-formance that might be realized, or an actual demand that might be met. These are then translated into requirements on further development. Resources can be dedicated to realize the requirements. After a first round of development work and evaluation of results, specific obstacles and/or specific expectations will be iden-tified and again translated into requirements on further work. Activities to address these requirements will often occur in a 'protected space', an environment which is relatively shielded from outside scrutiny. Expectations actually support the creation of such 'protected spaces'. This succession of promises, requirements, ongoing work, and more specific promises and related requirements (a promise-requirement cycle (Van Lente 1993)) continues until a working artefact or system is realized. It may be stopped prematurely if progress is disappointing.

Thus, the promise-requirement cycle creates a trajectory of development. At each turn of the cycle decisions have to be taken – necessarily based on expectations – whether to continue to explore or whether to be more focussed. Overall, the trajectory moves from requirement-constrained exploration of an option to a focus on exploitation (March 1991; Verganti 1999). By then, other perhaps also viable options might be neglected or become less important.[8]

Dual dynamics of promises

The two dynamics are not independent. Promise-requirement cycles can and will refer to the umbrella promise for legitimation, and the umbrella promise may want to draw on some successful realizations to remain credible. To capture such dual dynamics of promises, Figure 3.1 visualizes the dual dynamics. The basic

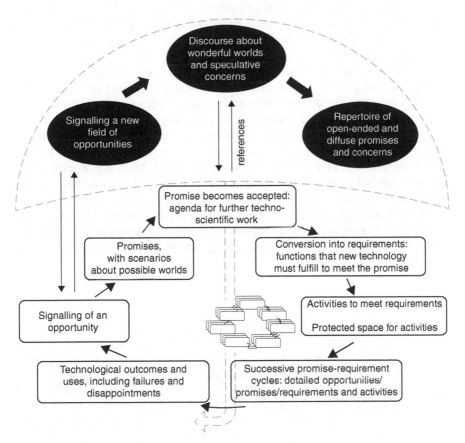

Figure 3.1 Dual dynamics of promises

coupling is how the signalling of a promising option, say a new nano-structured material, is linked to the signalling of a new field of opportunities like nanotechnology in general. The key question in further interactions is whether the success or failure of concrete projects is blamed on the umbrella promise or on circumstances (cf. Rip 2012).

Different patterns can emerge. We have summarized patterns in dual dynamics in Table 3.1 and highlighted the different role umbrella promises play in relation to more concrete promise-requirement cycles. We will zoom in on the pattern of 'waiting games', where the reluctance of actors to invest in concrete developments, while they continue to refer to the umbrella promise, has become like a rule of the game.

Waiting strategies

Before we discuss the pattern of a waiting game, we must consider waiting strategies of actors, as these can be followed without there being a waiting game. For example the recurring waiting strategy of big pharmaceutical companies to not invest in exploratory R&D and lateral innovation, relates to their sunk investments in the business model of blockbusters, and their power to buy into eventually interesting developments realized by small firms and/or public research groups, at a later stage when the promise is more definite.[9] It is a symbiosis game, and some start-up R&D companies in the sector actually say in their business model that their aim is to eventually sell out to a big pharmaceutical company.

Waiting strategies can be the result of firm-internal decisions estimating costs and benefits of taking different timing decisions.[10] For example postponing in order to avoid the costs of learning at an early stage, or conversely, obtaining a competitive advantage by getting ahead on the learning curve. For novel and possibly disruptive technologies, such assessments are difficult to make (Barnett 1990): there is lack of a common technological frame, there are no established networks, and no standards and infra-technology (Tassey 1991). The new product development literature (e.g. Mullins and Sutherland 1998) has emphasized that in transient and turbulent environments decision makers will inevitably have difficulty in estimating the amount of resources that should be committed in the face of market uncertainty as well as the timing with which such resource commitments should be made available.

Waiting games

In the situation of a waiting game, the reluctance of different actors to invest effort and money is not a matter of individual decisions which happen to gravitate towards little or no investing. There are mutual dependencies and informal rules which are being followed, so it is a game, played by real actors. '[A] game exists if courses of action are in fact interdependent, so that the outcome achieved will be affected by the choice of both (or all) the players' (Scharpf 1997, 7).

Specific for the waiting game linked to promises around NEST is the appeal of the umbrella promise which requires actors to pay some attention. Even in cases

Table 3.1 Different patterns in dual dynamics of promises of emerging technologies

	Hype-disappointment cycles	Promise icons	Priority setting	Waiting games
Dynamics	While actors are concerned about possible disappointment, exaggeration is perceived to be inevitable to create visibility and attract funding. Umbrella promises are hype-friendly.	Concrete technology developments refer back to umbrella promises which remain unchecked. Success in developments is attributed to the 'force' of the umbrella promise, while failure is related to other factors.	Umbrella promises invite priority setting and funding (related to grand challenges). Long-term goals are more important than concrete innovation.	Appealing umbrella promises keep actors 'in the game', while uncertainties about demand from users and uncertainties about specific technology directions make the interdependent actors reluctant to take concrete action.
Outcomes	Big umbrella promise eventually collapses, a shake-out occurs, work continues on the few technologies that show concrete promise.	Umbrella promises survive, despite failures in concrete technology developments which refer back to them.	With their partial institutionalization, umbrella promises provide stable backdrop for agenda-building and acquisition of funding.	Actors recognize umbrella promises, but will not take the risk of investing heavily in concrete developments.
Empirical examples	Hype-cycle introduced by Gartner Group for ICTs; hype-cycles as part of 'folk theories' about nanotechnology (Rip 2006)	Electronic super highway (Konrad 2004)	European Technology Platforms (ETPs) such as ETP Nanomedicine (this paper)	Impasses in OLAE and drug delivery (this paper)

where there is little involvement, as with big pharma in the area of nano-enabled drug delivery systems, they still want to closely monitor developments (as their researchers and strategy staff do anyway). So, actors choose not to exit the game. The umbrella promises remain on the radar of various actors even if little actual developments take place.

There are further features of this kind of waiting game, like the fact that producing and circulating open-ended promises need not require much dedicated investments (although that depends on the target audience). The repertoire of promises can remain alive without much effort.

More important is the combination of two types of uncertainties: at the development and supply side and at the user and demand side. At the user side there is little incentive to articulate specific demands (to be met by the developers/suppliers) because it is unclear in which direction the technology will develop, and which performance can actually be achieved. At the supply side, the uncertainty about demand creates reluctance to pro-actively choose for one or another development, even when it is recognized that one must at one moment shift from exploration to exploitation. There is no 'solution', other than one or another actor being willing to stick out its neck and take a risk. And this happens occasionally. Firms, for example in the ICT sector, have been willing to introduce products for which demand was still unclear. Users may commit themselves to accept a product to be developed, as when governments go for procurement to stimulate technological innovation (Edler and Georghiou 2007; Georghiou 2007).

Thus, waiting games may be overcome. Actors actually consider possibilities for this to happen and can discuss strategies in order to change what they perceive as an undesirable situation.[11] One way is coordination of emerging technologies (cf. the discourse of European Technology Platforms). Another way is the establishment of novel institutions to reduce uncertainties (Te Kulve 2010). Such moves themselves require investing in a 'collective good', and actors may well be reluctant to do so. In our empirical and scenario work we identified attempts by actors to overcome waiting games and saw how difficult it was for them to gather sufficient momentum to make a difference.

Case studies

The concept of waiting games, linked to the dual dynamics of promises, will structure our case studies of two domains of development in the broad field of nanotechnology.[12] Promises can be made for the broad field as a whole. In our two case studies we will focus on identifying the umbrella promises specific to each of the two domains, and trace how they function.

The second step in our analysis is to check for reluctance to invest in concrete developments, so that there will be few and/or only partial promise-requirement cycles. In a third step, we trace the constellation of mutual dependencies and thus specifics of the waiting game in that domain, and consider how actors are coping with waiting games.

The two cases are interestingly different. The drug delivery sector is an intersection between two specific value chains, drug development and formulation & device development, to serve an end market. The role of nanotechnology is enabling: it will improve what is done already, and only occasionally introduce new functionalities. OLAE on the other hand, as a new field of opportunities, can serve many different value chains, and the range of technological options, niche developments, and relevant actors is much larger. In the case of OLAE we see all the elements and dynamics of a waiting game linked to dual dynamics of promises. Nano-enabled targeted drug delivery on the other hand is not a new field, even if there are new options. Other interactions and other responses to the waiting game will be visible there, compared with the OLAE case.

Organic large area electronics

OLAE is a technology platform and as such opens up a new field of varied opportunities.[13] The discovery of organic semi-conducting materials (Shirakawa et al. 1977) led to visions about the use of conjugated organic molecules and organic/ inorganic composites to conduct current, emit light, and act as semiconductors, and to exploit the mechanical properties (flexibility) of these materials. This would create a shift from present silicon and lithography-based manufacturing in the electronics industry to potentially cleaner, more flexible and cost effective manufacturing processes like printing (Shaw and Seidler 2001). One effect might be that printer manufacturers and printing industry would make their entry in the electronics industry – a change in industry structure.

Since the 1970s the focus of activities has been on R&D to improve performance of the materials. Starting in the 1980s, effort was also invested in making useful working organic semiconductor devices. By now, there is recognition that complementary innovations are necessary, and further actors enter the world of OLAE. In addition to printing, new manufacturing technologies, process equipment, and encapsulation techniques have proven to be important.

The electronic and material properties of the new materials led to visions of a wide range of new products, but with little interaction with, or anticipation on, end users. At the moment, the domain of OLAE is still a world in which technology-linked relationships among actors dominate. Business to business relationships are explored, and 'new combinations' are formed across traditional value chains.

Umbrella promise

The umbrella term OLAE is used when linking the broad vision of technological opportunities with significant societal themes like global warming (alternative energy, higher efficiency devices), increasing mobility (light weight, increasing functionalities), and embedded intelligence (low cost sensing and RFID). This is visible in the text and title of the recent (2009) strategic research agenda of OLAE presented to the European Commission.[14] The rhetorically powerful title

'Towards green electronics' went together with reference to a potential market of $300 billion to be realized by 2027.

There are promises about the disruptive functionalities by authorities like Sir David King, former U.K. government chief scientific:

> In Britain we have a world-leading position in a technology that could wipe out silicon chip technology and could convert photovoltaics into easily accessible materials at a much cheaper price, and I am talking about plastic electronics.
>
> (House of Commons 2009, 1)

Current policy narratives are mobilized to create a sense of urgency, including the maintaining of a leadership position in the face of increasing competition:

> Organic and Large Area Electronics is one of the most promising areas in the field of electronics . . . the potential of this disruptive electronic area, which is predicted to reach a market size of about $10 billion in 2010 and grow exponentially thereafter until acquiring a market size equivalent to that of the silicon. . . . The cost-effectiveness of this technology for large area (unlike silicon) is acting as a catalyst. And the fact that the whole technology can be added on flexible plastic foils has also been a powerful driver for this field. . . . Concerning sustainable development, the advances could also be noticed in many areas. . . . Europe has a strong position and expertise in organic and large area electronics which derives from the number of academia institutions working in the area, ahead of East Asia and North America.
>
> (European Commission 2008, 14)

Industry analysts warn of the danger for Europe to lose the race (Harrop 2007). The chairman of IDTechEx highlighted this as a possible conclusion of IDTe-chEx's report '*Organic and Printed Electronics in Europe*':[15]

> Overall, Europe may be losing the race for the huge new business of printed electronics and the rejuvenation of society that it will bring. This is despite having far more academic institutions than East Asia working on the subject, the number being somewhat ahead of North America as well.

Such forecasts construct an economic interest in the field of OLAE, which is then replayed in policy documents. But there are also warnings about inflated expectations that cannot realistically be achieved, from influential consultancies like Cintelliq.[16]

The underlying assumption in this discourse of disruptive technologies (specified as printed RFID, cheap and disposable point-of-care devices, organic photovoltaics, and OLED for displays and lighting) is that progress along learning curves will be rapid enough to overtake existing technologies (cf. Christensen

1997). But this assumption remains unchecked, while one may be doubtful because of the complexities involved in crossing value chains, complementary technologies, and acceptance by users.

Reluctance to invest in product development

In our interviews and document analysis, the contrast was clear between hopeful research, conferences, and consortiums to mobilize support for OLAE, and the absence of concrete strategies and investment in product development. There is an abundance of options but no easy way to select for exploitation in actual product development. The conclusions of the Organic Electronics Association (2008, 5) emphasize this, when they say that it is not possible to define a 'Moore's Law' for organic electronics, or to define any killer applications – so there are no simple incentives to invest in specific product development.

Actors are concerned about lack of interest from end users.[17] At a stakeholder meeting organized by the European Commission in 2007, Wolfgang Clemens, Director of Organic Electronics Association (which represents more than 70 companies active in the field of OLAE), emphasized, '*End-users do not know enough about possibilities of the technology*' and proposed to educate end users (Clements, Mildner and Hecker 2007, 15). A competence matrix study on plastic electronics in the UK concluded: 'There is no incentive for companies to develop compatible standards or push for common technology platforms without a powerful end-user to insist on this' (King 2008).

The combination of lack of articulated demand (with end users, but also with business customers) and lack of articulated directions for product development creates a situation where actors are reluctant to invest. This is made explicit in an interview in an industry magazine with Martin Schmitt-Lewen, manager of functional printing at Heidelberg Company:

> 'I can't see Heidelberg commercially printing electronics or developing the equipment for a long while yet and there is no point in developing a printing kit or system when there are no existing customers ready to buy them, considering very few companies in the printed electronics market are scaling up production, particularly in RFID and active packaging space, there is no requirement for large print press systems'. He argues further in this interview 'We want to avoid speculatively developing equipment or printed electronics products until the technology and the market are more mature'.
>
> (Anonymous 2009, 37)

This is not just a risk averse approach. If it were, Heidelberg Company would not bother, and just exit. But it maintains a presence in the OLAE world, just in case. Many innovation actors do so, for example by being involved in public or semi-public projects, conferring symbolic support to R&D in the field. They may be reluctant to invest, but are also on the lookout for opportunities to do something.

Constellations in the waiting game

There are actual and projected mutual dependencies which shape strategies and patterns of interaction. Upstream actors like materials suppliers, printer manufacturers and ink producers are a key example.

Materials are crucial for device performance. Exploring the qualities of specific ink-jet techniques requires printer manufacturers to form new alignments with material suppliers and ink manufacturers (Cleland 2003, 108). But there are conflicting business models.

The use of materials in the printing technologies for the production of a printed OLED or organic transistor is much less than in the traditional deposition technologies. Chemical suppliers are used to sell large volumes of materials, but in the new situation they must adapt their pricing strategies.[18] They must also learn to produce for niche applications of organic electronics. Chemical suppliers may not make these moves of their own accord.

Uncertainty about value capture and profitability, a general issue, is felt in the printing industry. At present, the ink-jet printing industry is profitable because of the sales of ink supplies rather than of printers.[19] The industry may want to become pro-active as existing patents on OLED materials will soon expire.[20] Ink-jet companies may then leverage economic advantage by formulating own materials or partnering with firms that have the capability to do so (Cleland 2003, 108). Or new entities may emerge in the value chain, like design houses for products with different specifications. But this is a realistic option only when a global capacity for dedicated production is in place and some standardization has occurred.[21] So the waiting game may strike again: there are no incentives to invest in building up such a capacity and creating standardization.

While there is a lot of work in the field of organic electronics on improvement of materials' performance, requirements derived from downstream applications get less attention. One such challenge is encapsulation, a critical problem in realizing organic electronics products because organic materials are highly sensitive and fragile and deteriorate when exposed to oxygen and moisture (Patterson 2009, 160–163). Lack of adequate encapsulation techniques and materials constitutes a reverse salient for the further expansion of OLAE.[22] Dedicated work on the reverse salient is like producing a collective good, and return on such an investment will materialize only when the expansion of OLAE applications is achieved. To overcome this version of the waiting game, action at the collective level may well be necessary.

This diagnosis of the situation was actually taken up in one of the scenarios we created for further development of OLAE and its embedding in society.[23] We had the UK government initiate a procurement policy to give OLAE a boost: development of new RFID-based electronic passports and guaranteeing their use (this is procurement policy). Some other countries sensitive to terrorism joined the initiative. Eventually, the technical performance was insufficient to guarantee privacy, and the project was stopped. But the challenge of encapsulation had to be met, and effective materials and techniques were now available generally. Others could profit, for example in the government-supported development of organic photo-voltaics in Germany. Insofar there is a moral to this scenario, it is that collective-level actors can play an important role, but that the challenge of the reverse salient of encapsulation was (partially) resolved as a by-effect of a procurement project, rather than addressed as such.

Nano-enabled drug delivery systems

Drug delivery systems, including nanotechnology-enabled systems, are part of the formulation stage of drug manufacturing. A drug delivery system is a formulation or device 'that delivers therapeutic agent(s) to desired body location(s) and/or provides timely release of therapeutic agent(s). The system, on its own, is not a therapy, but improves the efficacy and/or safety of the therapeutic agent(s) that it carries'.[24]

The drug delivery sector is an intersection of two product value chains, drugs and delivery. The value chain of drug development involves large pharmaceutical companies that have their focus on developing new chemical entities rather than novel formulations (Bawa 2007; Harris et al. 2004); the latter are outsourced to drug delivery product companies, cf. De Leeuw, de Wolf and van den Bosch (2003). The drug delivery sector was dominated by the customer-oriented pharmaceutical companies, but the new interest in carriers has enabled drug delivery companies to become somewhat independent of the pharmaceutical world.[25] Nanotechnology enables this sector to do better, and is not seen as disruptive: it can create new functionalities, but basically focuses on maximizing drug effectiveness through controlled release and accuracy in reaching targets.

The increased academic interest in nanotechnology-enabled drug delivery is visible in the steep rise of research papers and patents since the early 2000s (Kim et al. 2009). The rise in research funding for the combination of nanotechnology and drug delivery, predicated on the promises of nanotechnology, has given the field of drug delivery research a new impulse (cf. Boyd 2008). With its focus on coordination, the European Technology Platform (ETP) Nanomedicine actively promotes developments of nano-enabled drug delivery systems.

Umbrella promise

The 2005 vision paper of the ETP Nanomedicine sets out the promises, starting with the general promise of nanotechnology including the promise of nanoparticles to reach their targets as if they were a magic bullet.[26] They characterize nanotechnology as a crucial enabler to reach goals in the medical and health care sector:

> The long-term objective of drug delivery systems is the ability to target selected cells and/or receptors within the body. At present, the development of new drug delivery techniques is driven by the need on the one hand to more effectively target drugs to the site of disease, to increase patient acceptability

and reduce healthcare costs; and on the other hand, to identify novel ways to deliver new classes of pharmaceuticals that cannot be effectively delivered by conventional means. Nanotechnology is critical in reaching these goals.

(p. 8)

At the same time, a sense of urgency is created by referring to the narrative of competition (e.g. 'Europe has a strong position'; the need to 'boost this promising field'), leading to a call for close collaboration in research activities in the field of nanomedicine:

At present Europe has a strong position in the emerging field of Nano-Medicine that has a high potential for technological and conceptual break-throughs, innovation and creation of employment. NanoMedicine is an area that would benefit from coordination at European level. Thus, close coop-eration between industry, research centres, academia, hospitals, regulatory bodies, funding agencies, patient organisations, investors and other stake-holders could dramatically boost this promising field. In response to these challenges, scientific experts from industry, research centres and academia convened to prepare the present vision document regarding future research priorities in NanoMedicine.

(p. 5)

The dual dynamics pattern of 'priority-setting' is visible here. The umbrella promise is referred to widely in the literature and in declarations at conferences and in the media. Nanotechnologies are considered to have 'unique qualities' (Emerich and Thanos 2006) and provide 'extraordinary opportunities' to con-tribute to targeting problems for major diseases such as cancer therapies (Ferrari 2005). In addition, nanotechnology-enabled drug delivery technologies are also expected to contribute to finding medicine to treat other diseases such as infec-tions, metabolic, and auto-immune diseases (Couvreur and Vauthier 2006).

While the general promise is recognized, and considered important, there are also sceptical comments on how far the nanotechnology-enabled targeting has actually advanced (Ruenraroengsak, Cook and Florence 2010), and whether exist-ing carriers and approaches might be good enough already. At conferences and in interviews (by one of us, HtK), nanotechnologies were often considered to be not very attractive to develop in-house. Promises that nano-enabled drug delivery might extend the life-cycle of drugs (by allowing further patenting) were met with scepticism. In terms of performance, the promised improvement of existing medicines, e.g. by increasing bioavailability or improving of the dosage regime, might not be so attractive, as therapeutical gains were expected to be limited and nanotechnologies would have to compete with well-developed conventional approaches. Delivery of new active compounds such as biopharmaceutical enti-ties and disease targeting were considered to be more promising (Wagner et al. 2006; Keller 2007; Boyd 2008).

Reluctance to invest in product development

In spite of the general need for new pharmaceutical technologies by pharmaceutical companies, they do not regularly involve themselves in nano-enabled drug delivery research projects or in broad stakeholder forums such as the ETP Nanomedicine.[27] Researchers working on nanotechnology-enabled drug delivery technologies complain about this lack of interest. The EU-funded consortium MediTrans is explicit in its diagnosis how this impedes the introduction of nano-technologies in the clinic:

> It is now well known that a reliable targeting system is essential for successful drug delivery in many serious disease situations. It is becoming increasingly recognized that a major limitation, impeding the entry of targeted delivery systems into the clinic, is that new concepts and innovative research ideas within academia are not being developed and exploited in collaboration with the pharmaceutical industry. Thus, an integrated 'bench-to-clinic' approach realized within a structural collaboration between industry and academia, is required to safeguard and promote the progression of targeted nanomedicines towards clinical application.
>
> (MEDITRANS 2007a)

While big pharmaceutical companies are interested in innovative medicines,[28] they do not invest in nanotechnology-related research. Director Keller from GlaxoSmithKline mentioned during a conference on Medical Nanotechnologies that they 'recently had shut down their department as it was not the time to create a nano department' (Keller 2007).

There appear to be two waiting strategies involved. The main one derives from the overall promise of nanotechnology and assessment of the more specific promises about nano-enabled targeted drug delivery, as convincing enough to start to invest. This is then compounded by big pharma's focus on drugs, i.e. active compounds rather than formulations and or delivery systems. And it is linked to their general waiting strategy: let others take the risk of novel approaches. Big pharmaceutical companies such as Glaxosmithkline and AstraZeneca preferred to co-operate with other parties rather than doing development themselves (Keller 2007; Washington 2007).[29]

In their review of technological and commercial developments, Couvreur and Vauthier (2006) pointed out that a 'confident climate' for new nanosystems has to be developed on the basis of already positive experiences (both clinically and in regulatory terms) of such systems. But they also observed reluctance of large pharmaceutical companies towards nano drug delivery: 'Today, most developments are carried on by small entrepreneurial firms including many spin-ups that cannot support themselves as yet on current revenues, whereas big pharmaceutical companies seem still awaiting for more successes'. Wagner et al. (2006) speculated that firms were waiting for the first nano blockbuster before they would initiate heavy investments in this area.

In response to this waiting game, researchers, and other promoters of nano-medicine in general, and nano-enabled targeted drug delivery in particular, moved from general promises to more specific ones, in the hope that this would lead to investment in actual development. This is visible in the 2009 Roadmap Document of the ETP Nanomedicine. Recommendations were offered, and to persuade industry there was a veiled reference to a possible loss of competitive position in the global market:

> The Strategic Research Agenda of the ETP Nanomedicine was drafted in 2006 with a broad range of options highlighted. Over the intervening years it has become increasingly clear to the industrial sector that an academic driven or laissez-faire approach to Nanomedicines will be an inefficient process. It is recognized that it is now time to make more detailed specific recommendations. . . . Successful translation of research results from academia into products has been identified to be one of the major challenges in this innovative science based area. Strategies to foster and initiate this translatability must be developed and implemented to help European research and industries remain competitive in the global market.
>
> (ETP Nanomedicine 2009, p. 4)

There continues to be support for R&D, and public interest has not disappeared, but actual product development remains limited (Te Kulve 2011).

Constellations in the waiting game

There might be symbiotic relationships: '[S]maller companies and start-ups rely on partnering with big pharma that provide funding for the clinical trials' (Wagner et al. 2006, 36). The involvement of big pharma would bring in knowledge and experience with respect to development and regulatory processes (Eaton 2007). Not much is happening in practice, also because of uncertainties about intellectual property rights.

A key aspect of the constellation is the regulatory set-up. Regulation of new drugs and drug delivery systems has well-articulated procedures, and for new drugs quite elaborate and costly. Because new drugs are approved after going through pre-clinical and clinical tests, this also puts constraints on what can then be changed in terms of formulation without doing further studies (Washington 2007). For nano-enabled drug delivery systems which bridge drug, device, and formulation development, it is not clear how regulatory processes will work out, also because of uncertainties about standards and methodologies for assessments of new nano-enabled technologies. Authorities will wait until they know which technologies are under development before checking whether existing procedures are adequate, or modifications are in order. Drug delivery and pharmaceutical companies will not develop new products if there is no assurance about regulatory requirements.

At the moment, the key challenge is the gap between the enthusiasm of researchers and other promoters of nano-enabled targeted drug delivery systems, justified in terms of umbrella promises, and the reluctance of clinical actors and pharmaceutical companies to involve themselves in the development of such systems. One visible strategy from the side of the promoters is to shift from pushing their 'magic bullets' to emphasize the need for translational research (Te Kulve 2011). In doing this, they relate to a general move in this direction in nanomedicine, and in pharmaceutical research generally.

Possible next steps were explored in scenarios.[30] A key move to break through the waiting game was coordination through platforms and consortia together with other actors in the value chain, up to patient organizations. Depending on how the interventions were received and had to be modified, different (and contested) development paths were pursued. One route was to go for the 'safe' option of liposomes and other biodegradable carriers. Cancer therapies, another route and one which had captured media attention, turned out to be quite complex, and lead to niche applications at best.

In conclusion

The case of nano-enabled targeted drug delivery is interesting because the waiting game linked to dual dynamics of promises is additional to the general waiting strategies of big pharma with respect to novel approaches. The case of OLAE is interesting because of the potentially disruptive character of OLAE, so that actors are at a loss which directions to take. Thus they fall back on the safe option (in the short term) to wait and see.

A key feature of the two cases, and of NEST in general, is the combination of uncertainties about eventual performance and uncertainties about customer/user demands and requirements. There attempts to reduce uncertainties through new alliances and consortia, but the basic response of actors was reluctance to invest in concrete product development. The overall promises of OLAE and (to a lesser extent for some of the actors) of nano-enabled drug delivery were recognized, and in fact there was a quasi-certainty about them. OLAE will arrive, unavoidably so, even if 'we' have to put a lot of effort into realizing the promise. This is what turned the situation into a waiting game: exiting the game, by forgetting about OLAE, is not an option.

If there are waiting games, there will be less development, and thus less learning about possibilities than could have been the case. On the other hand, the pause introduced by waiting games may become an occasion to reflect on desirable directions and alternatives, rather than falling for the sense of urgency induced by the umbrella promise.

Waiting games around NEST appear to be unavoidable. But innovation and policy actors driven by their sense of urgency will see this situation as undesirable and will try to overcome the waiting game.[31] They may not be successful, because actions at the collective level are necessary to break through waiting games, and these are difficult to achieve. A small step in the direction of doing better is provided by our scenario exercises and their discussion in strategy-articulation workshops with different stakeholders.

Thus, what we have done in this paper is more than observing and documenting a pattern, the pattern of waiting games that keep actors involved in the domain while this does not lead to much investment in concrete developments. We added an explanation in terms of dual dynamics of promising, showed that these dynamics were actually occurring, and explored some of the actual variety in the two cases that we presented. Since actors recognized there were waiting games (even if they need not think of explanations), we could also ask how they were coping, and perhaps trying to overcome the waiting games. Here, our understanding of constellations of actors and mutual dependencies could come in to help them become more effective, or at least, more reflexive.

Notes

1 The term 'halo' (around nanotechnology) was introduced by Vincent Bontems in a paper, 'How to accommodate to the invisible? Nano-impressionism', contributed to the workshop on 'Imag(in)ing the Nanoscale. Interactions between science, public media, and art', Bergen (Norway), 27–29 January 2011.
2 Ruef and Markard analysed such dual dynamics for one area of fuel cell development; they used the term 'frame' for what we called 'big but open-ended promises'. (Ruef and Markard 2010)
3 In their analysis of economics of techno-scientific promises, Joly, Rip and Callon (2010) discuss master narratives of promise, and how this drives policies, but do not recognize how this, paradoxically, can lead to waiting games.
4 Compare: 'The calculative agents on the idealized market of economic theory will identify possible "states of the world", rank them, and define choices and actions in those terms' (Callon 1998, 4) – which then change the current 'state of the world'.
5 The EC created ETPs as a general instrument in order 'to bring together all interested stakeholders to develop a long-term shared vision, create roadmaps, secure long-term financing and realize a coherent approach to governance' (European Commission 2004, 10).
6 This point is emphasized by Rip and Voss (2011) in their analysis of umbrella terms like 'nanotechnology' and 'sustainability' in science and science policy.
7 Rip (2011) signalled a funding race (between states) for nanotechnology, rather than innovation races.
8 Van Merkerk and Robinson (2006) have discussed this as an example of emerging irreversibility, and linked it with the phenomenon of path dependency because of sunk investments.
9 Cf. Wagner et al. (2006, 36) highlighting big pharma's reluctance to invest in novel technologies in the field of biopharmaceuticals and nanomedicine.
10 See the seminal work by McDonald and Siegel (1986) and Pindyck (1991) and the literature on irreversible investments. Adoption (e.g. going for a particular development) can be viewed as a strategic switching-time decision problem for agents facing an ongoing stochastic operating benefit, plus sunk investment costs (Moretto 2000).
11 Ironically, calls of civil society groups for moratoria on the introduction of nanotechnologies like nanoparticles until their safety is ascertained take the waiting game as a desirable situation.
12 The case material is based on extensive and detailed study of documents, interviews, observations which have been reported in Te Kulve (2011) and Parandian (2012). The diagnosis of waiting games was used to create scenarios showing repercussions of one or another attempt to overcome the waiting game, which were subsequently used in

interactive strategy-articulation workshops with stakeholders in the domain (Te Kulve 2011; Parandian 2012).

13 The definitions of the new field have not stabilized and different adjectives are used, e.g. flexible, organic, large area, plastic, and polymer electronics. The term OLAE is used more generally, as an umbrella acronym, to refer to a field of innovative opportunities of the various technologies and their possible applications.

14 Strategic Research Agenda Organic & Large Area Electronics. 2009. Final Version 1.4. Last retrieved 3 May 2011. www.photonics21.org/download/olae_sra.pdf

15 This IDTechEx report analyses and compares the activities of 248 European organizations in the sector.

16 Cintelliq is an information service and technology consulting company for organic semiconductors.

17 This was emphasized by Tom Taylor, CEO of PETEC (UK), during a seminar on plastic electronics in May 2009 organized by the IET Microsystems and Nanotechnology Network and the Institute of Nanotechnology in London. Also in an interview with Michael Butler of Unilever (UK), with one of us (AP).

18 As Eliav Haskal director of lighting strategic partnership in Philips Company, noted (in an interview with one of us, AP), for the short term they can charge premium prices to make up for the R&D costs they make hoping to supply to mass markets in the future.

19 This point was emphasized by Hylke Veenstra, of OCE The Netherlands, during his presentation at the Conference 'Tomorrow's Electronics', Utrecht (The Netherlands), May 2009, and in an interview with one of us (AP).

20 Interview (by AP) with Professor Paul Blom- University of Groningen and Holst Centre, Eindhoven.

21 Interview (by AP) with Professor Andres Dietzel- Eindhoven University and Holst Centre, Eindhoven.

22 A reverse salient is a subsystem that blocks further expansion of the entire system, so that improving other subsystems will not make a difference as long as the reverse salient is not overcome (Hughes 1983; Ortt and Dedehayi 2010).

23 Three scenarios were created to support interactive strategy articulation workshops with stakeholders about OLAE, one in Eindhoven (supported by Plastic Electronics Foundation) and a second one in Heidelberg (supported by InnovationLab GmbH). See Robinson (2009) for the general approach of such scenarios.

24 From www.drugdel.com/glossbot.htm

25 Delivery devices can not only be used as carriers for drugs but can also be applied for diagnostic (imaging) purposes and in the context of food, as carriers for nutraceuticals.

26 The notion of a 'magic bullet' (Zauberkugel) was introduced by Paul Ehrlich, more than a century ago, to characterize the nature of the newly developed chemical drugs.

27 Whereas the vision paper was signed by high-level managers of companies such as GlaxoSmithKline and Schering, no big pharmaceutical companies were present within the drug delivery working group of the ETP Nanomedicine.

28 A European Technology Initiative 'Innovative Medicines' has been created, co-ordinated by the European Federation of Pharmaceutical Industries and Associations (EFPIA). It eventually developed into a Joint Technology Initiative in which there was strong participation by large pharmaceutical companies (Innovative Medicines Initiative 2008).

29 Observations by one of us (HtK) during the conference Investing in Medical Nanotechnologies II, London 2007. See also Earl (2007). 'Investing in Medical Nanotechnologies II – Review.' Retrieved July 15, 2009, from http://nanotechweek.blogspot.com/2007/12/investing-in-medical-nanotechnologies.html.

30 Three scenarios based on different interventions to overcome the waiting game were developed and discussed in a strategy-articulation workshop with stakeholders, January 2010, organized in collaboration with branch organizations Nefarma and NIABA.

31 In terms of Hirschman (1970), this is the 'voice' rather than the 'exit' option.

References

Anonymous. 2009. Between the Lines: What Is Heidelberg's Real Plan for Printed Electronics? *Plastic Electronics Magazine*, 1(6): 32–39.

Barnett, W.P. 1990. The Organizational Ecology of a Technological System. *Administrative Science Quarterly*, 35(1): 31–60.

Bawa, R. 2007. Patents and Nanomedicine. *Nanomedicine*, 2(3): 351–374.

Borup, M., N. Brown, K. Konrad, and H. Van Lente. 2006. The Sociology of Expectations in Science and Technology. *Technology Analysis and Strategic Management*, 18: 285–298.

Boyd, B.J. 2008. Past and Future Evolution in Colloidal Drug Delivery Systems. *Expert Opinion Drug Delivery*, 5(1): 69–85.

Brown, N., and M. Michael. 2003. A Sociology of Expectations: Retrospecting Prospects and Prospecting Retrospects. *Technology Analysis & Strategic Management*, 15(1): 4–18.

Brown, N., B. Rappert, and A. Webster. 2000. *Contested Futures: A Sociology of Prospective Techno-Science*. Aldershot: Ashgate.

Callon, M., ed. 1998. *The Laws of the Markets*. Oxford, Malden: Blackwell Publishers.

Christensen, C.M. 1997. *The Innovator's Dilemma: When New Technologies Cause Great Firms to Fail*. Cambridge, MA: Harvard Business Press.

Cleland, T.A. 2003. *Printed Electronics: The Next Inkjet Revolution?* Master Dissertation, Cambridge, MA: Sloan School of Management at MIT.

Clements, W., W. Mildner, and K. Hecker. 2007. *WG2: Roadmap, Barriers for Market Entry*. Coordination: OE-A. Presentation at Organics and Large Area Electronics Stakeholder Meeting, EU Commission, Brussels, 1 October 2007. Retrieved from ftp://ftp.cordis.europa.eu/pub/fp7/ict/docs/events-3-20071001-wg2-activities-presentation_en.pdf

Couvreur, P., and C. Vauthier. 2006. Nanotechnology: Intelligent Design to Treat Complex Disease. *Pharmaceutical Research*, 23(7): 1417–1450.

De Leeuw, B.J., P. de Wolf, and F.A.J. van den Bosch. 2003. The Changing Role of Technology Suppliers in the Pharmaceutical Industry: The Case of Drug Delivery Companies. *International Journal of Technology Management*, 25(3/4): 350–362.

Earl, P.M. 2007. *Investing in Medical Nanotechnologies II – Review*. Retrieved July 15, 2009, from http://nanotechweek.blogspot.com/2007/12/investing-in-medical-nanotechnologies.html.

Eaton, M. 2007. Nanomedicine: Industry-Wise Research. *Nature Materials*, 6: 251–253.

Edler, J., and L. Georghiou. 2007. Public Procurement and Innovation: Resurrecting the Demand Side. *Research Policy*, 36(7): 949–963.

Emerich, D.F., and C. Thanos. 2006. The Pinpoint Promise of Nanoparticle-based Drug Delivery and Molecular Diagnosis. *Biomolecular Engineering*, 23: 171–184.

ETP Nanomedicine. 2009. *Joint European Commission/ETP Nanomedicine Expert Report 2009: Roadmaps in Nanomedicine Towards 2020*.

European Commission. 2004. *Communication From the Commission: Towards a European Strategy for Nanotechnology*. Luxembourg: Office for Official Publications of the European Communities.

European Commission. 2008. *Project Portfolio Sixth and Seventh Research and Development Framework Programmes: Organic and Large Area Electronics*. EU Commission, DG Information Society and Media. Retrieved from ftp://ftp.cordis.europa.eu/pub/fp7/ict/docs/organic-elec-visual-display/olae-project-portfolio_en.pdf

Ferrari, M. 2005. Cancer Nanotechnology: Opportunities and Challenges. *Nature Reviews Cancer*, 5: 161–171.

Geels, F., and W. Smit. 2000. Lessons From Failed Technology Futures: Potholes in the Road to the Future. In *Contested Futures: A Sociology of Prospective Techno-Science*, eds. Brown, N., B. Rappert, and A. Webster. Aldershot: Ashgate.

Georghiou, L. 2007. *Demanding Innovation: Lead Markets, Public Procurement and Innovation, NESTA Provocation 2.* London: National Endowment for Science, Technology and the Arts, February 2007.

Harris, D., K. Hermann, R. Bawa, J.T. Cleveland, and S. O'Neill. 2004. Strategies for Resolving Patent Disputes Over Nanoparticle Drug Delivery Systems. *Nanotechnology Law & Business*, 1(4): N.A.

Harrop, P. 2007. *Is Europe Losing the Race?* IDTechEx. Retrieved May 2011, from Printed Electronics World. www.printedelectronicsworld.com/articles/is_europe_losing_the_race_00000669.asp?sessionid=1

Hirschman, A.O. 1970. *Exit, Voice, and Loyalty: Responses to Decline in Firms, Organizations, and States.* Cambridge, MA: Harvard University Press.

House of Commons. 2009. *Innovation, Universities, Science and Skills Committee-Fourth Report: Engineering: Turning Ideas into Reality. Part 3, Plastic Electronics Engineering: Innovation and Commercialization.* House of Commons, UK. Retrieved from www.publications.parliament.uk/pa/cm200809/cmselect/cmdius/50/5006.htm#note74

Hughes, T.P. 1983. *Networks of Power: Electrification in Western Society, 1880–1930.* Baltimore: John Hopkins University Press.

Innovative Medicines Initiative. 2008. *Boosting Biomedical Research Across Europe. Kick-Off of a New Public-Private Partnership for Research Funding.* Retrieved June 22, 2010, from http://imi.europe.eu/docs/press-release-03032008_en.pdf

Joly, P.B., A. Rip, and M. Callon. 2010. Reinventing Innovation. In *The Governance of Innovation: Firms, Clusters and Institutions in a Changing Setting*, eds. Arentsen, M.J., W. Van Rossum, and A.E. Steenge, 19–32. Cheltenham, UK: Edward Elgar Publishing.

Keller, T. 2007. *Nanotechnology: Cutting through the Hype: A Realistic Business Case for Pharma.* Investing in Medical Nanotechnologies II, Royal College of Surgeons, London, 28–29 November.

Kim, S., I.K. Kwon, I.C. Kwon, and K. Park. 2009. Nanotechnology in Drug Delivery: Past, Present, and Future. In *Nanotechnology in Drug Delivery*, eds. De Villiers, M.M., P. Aramwit, and G.S. Kwon, 581–596. New York: Springer Verlag.

King, Z. 2008. *Plastic Electronics in the UK.* (Lack of End-Users Is Likely to Constrain Access to Markets for UK SMEs). Retrieved from www.printedelectronics.net/PlasticElectronicsintheUK.htm#1

Konrad, K. 2004. *Prägende Erwartungen. Szenarien als Schrittmacher der Technikentwicklung.* Berlin: edition sigma.

Konrad, K. 2006. The Social Dynamics of Expectations: The Interaction of Collective and Actor Specific Expectations on Electronic Commerce and Interactive Television. *Technology Analysis & Strategic Management* 18(3): 429–444.

Lund Declaration. 2009. *Europe Must Focus on the Grand Challenges of Our Time, One Result of the European Research Conference 'New Worlds – New Solutions'.* Organized Under the Swedish Presidency, Lund, 7–8 July 2009. Retrieved from www.se2009.eu/en/meetings_news/2009/7/8/declaration_from_the_research_conference_in_lund_euro pean_research_must_focus_on_the_grand_challenges

March, J.G. 1991. Exploration and Exploitation in Organizational Learning. *Organization Science*, 2(1): 71–87.

McDonald, R., and D.R. Siegel. 1986. The Value of Waiting to Invest. *The Quarterly Journal of Economics*, 101(4): 707–728.

MEDITRANS. 2007a. *State of the Art*. Retrieved June 23, 2010, from http://web.archive. org/web/20070504221129/www.meditrans-ip.net/State-of-the-art.html

Moretto, M. 2000. Irreversible Investment with Uncertainty and Strategic Behavior. *Economic Modelling*, 17: 589–617.

Mullins, J.W., and D.J. Sutherland. 1998. New Product Development in Rapidly Changing Markets: An Exploratory Study. *Journal of Product Innovation Management*, 15: 224–236.

Organic Electronics Association. 2008. *White Paper OE-A Roadmap for Organic and Printed Electronics* (May 2008 version). German Engineering Federation (VDMA). Retrieved from www.novaled.com/downloadcenter/oe-a_roadmap_white_paper_2008_may_public.pdf

Ortt, J., and O. Dedehayi. 2010. *Factors Hampering Technology System Progress During the Life Cycle: What Are These Factors and How to Deal With Them?* International Conference on Management of Technology, Cairo (Egypt) 2010, Proceedings.

Parandian, A. 2012. *Constructive TA of Newly Emerging Technologies: Stimulating Learning by Anticipation through Bridging Events*. PhD Thesis, Delft: Technical University of Delft.

Patterson, P. 2009. *Beyond the Beaker, How to Achieve Successful Market Adoption for Emerging Technologies*. Beaverton, OR: Kohritsu Press.

Pindyck, R.S. 1991. Irreversibility, Uncertainty, and Investment. *Journal of Economic Literature*, 29: 1110–1152.

Rip, A. 2011. Science Institutions and Grand Challenges of Society: A Scenario. *Asian Research Policy*, 2(1): 1–9.

Rip, A. 2012. Contexts of Innovation Journeys. *Creativity and Innovation Management*, 21(2): 158–170.

Rip, A., and J.W. Schot. 2002. Identifying Loci for Influencing the Dynamics of Technological Development. In *Shaping Technology, Guiding Policy*, eds. Williams, R., and K. Sørensen, 158–176. Cheltenham: Edward Elgar.

Rip, A., and J.-P. Voss. 2011. *Umbrella Terms in the Governance of Emerging Science and Technology: A Nexus Between Science and Society?* Enschede/Berlin: University of Twente/Technical Universitry of Berlin. Manuscript.

Robinson, D.K.R. 2009. Co-Evolutionary Scenarios: An Application to Prospecting Futures of the Responsible Development of Nanotechnology. *Technological Forecasting and Social Change*, 76(9): 1222–1239.

Ruef, A., and J. Markard. 2010. What Happens After a Hype? How Changing Expectations Affected Innovation Activities in the Case of Stationary Fuel Cells. *Technology Analysis & Strategic Management*, 22(3): 317–338.

Ruenraroengsak, P., J.M. Cook, and A.T. Florence. 2010. Nanosystem Drug Targeting: Facing Up to Complex Realities. *Journal of Controlled Release*, 141: 265–276.

Scharpf, F.W. 1997. *Games Real Actors Play: Actor-Centered Institutionalism in Policy Research*, ed. P. Sabatier, Theoretical Lenses on Public Policy. Boulder, CO and Oxford: Westview Press.

Shaw, J.M., and P.F. Seidler. 2001. Organic Electronics: Introduction. *IBM Journal of Research and Development*, 45(1): 5.

Shirakawa, H., E.J. Louis, A.G, MacDiarmid, C.K. Chiang, and A.J. Heeger. 1977. Synthesis of Electrically Conducting Organic Polymers: Halogen Derivatives of Polyacetylene, (CH)x. *Journal of the Chemical Society, Chemical Communication*, 578.

Tassey, G. 1991. The Functions Technological Infrastructure in a Competitive Economy. *Research Policy*, 20(4): 345–361.

Te Kulve, H. 2010. Emerging Technologies and Waiting Games: Institutional Entrepreneurs around Nanotechnology in the Food Packaging Sector. *Science, Technology & Innovation Studies*, 6(1): 7–31.

Te Kulve, H. 2011. *Anticipatory Interventions and the Co-Evolution of Nanotechnology and Society*. PhD Dissertation, Enschede: University of Twente.

Van Lente, H. (1993). *Promising Technology*. PhD Thesis, Enschede: University of Twente.

Van Lente, H., and A. Rip. 1998. Expectations in Technological Developments: An Example of Prospective Structures to be Filled in by Agency. In *Getting New Technologies Together: Studies in Making Sociotechnical Order*, eds. Disco, C., and B. Van der Meulen, 203–231. Berlin: Walter de Gruyter.

Van Merkerk, R.O., and D.K.R. Robinson. 2006. Characterizing the Emergence of a Technological Field: Expectations, Agendas and Networks in Lab-on-a-Chip Technologies. *Technology Analysis & Strategic Management*, 18(3/4): 411–428.

Verganti, R. 1999. Planned Flexibility: Linking Anticipation and Reaction in Product Development Projects. *Journal of Product Innovation Management*, 16: 363–376.

Wagner, V., B. Hüsing, S. Gaisser, and A.-K. Bock. 2006. *Nanomedicine: Drivers for Development and Possible Impacts*. Seville: Institute for Prospective Technology Studies.

Washington, C. 2007. *Medical Nanotechnology in the Pharmaceutical Industry: Opportunities and Problems*. Investing in Medical Nanotechnologies II, Royal College of Surgeons, London, 28–29 November.

4   Folk theories of nanotechnologists

Arie Rip

Published as

Folk Theories of Nanotechnologists, *Science as Culture 15*(4) (December 2006) 349–365

The world of nanotechnology is full of folk theories. Actors attempt to capture patterns in what is happening and be reflexive about them, so as to do better the next time. A clear example is the diagnosis of an impasse in agro-food GM technologies, coupled to statements about the need to avoid such impasses with nanotechnology. Since there is a claim that such patterns will recur (if we don't change our ways), there is generalization involved, and one can speak of a theory, here about public reactions to new technology. Calling it a *folk* theory implies that it evolves in ongoing practices, and serves the purposes of the members of the various practices. In this example, these would be nanotechnologists, policy makers for nanotechnology, other scientists and technologists, science watchers and commentators. What characterizes folk theories is that they provide orientation for future action.

There is a large and growing literature on folk sciences. For example Douglas and Atran (1999) depict forms of 'folk biology' as commonly accepted taxonomic categorizations of ecological systems (see also Atran, 1998). In the same way, 'folk physics' and 'folk chemistry' describe taken-for-granted categorizations and terminology that explain the physical, natural, and chemical world and help orient actors (Heintz et al., 2007, forthcoming). Just as there is folk physics and folk psychology, there is also folk sociology. It is part and parcel of the repertoires of everyday life. Elements of folk sociology can be articulated as such, and serve as folk theory.

Folk theories can be more or less explicit (this also depends on whether they are challenged or not). They are a form of expectations, based in some experience, but not necessarily systematically checked. Their robustness derives from their being generally accepted, and thus part of a repertoire current in a group or in our culture more generally (Swidler, 1986; Rip and Talma, 1998).

While ways to acquire knowledge about the social world is not principally different for such folk theories and for social science, actors as well as analysts attempt to capture patterns in what is happening, actors tend to short-circuit their analysis so as to serve their action perspective. They may then remain captured by their own folk theory: it colours their views of the world, and they interpret whatever they encounter in terms of this theory. In that sense, folk theories are conservative. They become cognitive frames (Lakoff, 1987), almost like folkways (Sumner, 1907). The important difference with folkways is that folk theories have content, and can be taken up in terms of their content.

Analysts asking similar questions, e.g. about lessons from GM, can critically evaluate folk theories of actors. Even while their own theories may be folk theories as well, be it for the different (and more critical) folk constituted by their colleagues.[1] Analysts can try to enlighten actors, even while often the actors are not keen to listen to analysts. This experience raises a further, reflexive question: how do such theories come about and why can they be so tenacious?

As a sociologist, one can try to explain why actors would develop and hold particular folk theories. Mary Douglas's cultural theory (based on her anthropological studies) does that, and quite effectively for situations of uncertainty and novelty (Douglas, 1982). Such an explanation is not deconstruction of the value of the theories, even if the actors might experience it that way. It is embedding the theories in the practices they function in anyway. Such positioning of the folk theories creates some distance and might lead to reflection, and what one might call sociological enlightenment. Such enlightenment can then make it easier to replace earlier folk theories by better theories. An example would be the criticism of the recurrent deficit model of the public, and replacing it with a public engagement model – the action perspective is recognizable, also at the side of the analysts.

I shall come back to these reflexive issues in the conclusion, because there appear to be interesting openings in the nano-world for sociological enlightenment. In the body of the paper, I present clusters of folk theories that are current in the nano-world, indicating why they must be positioned as folk theories rather than entering into detailed discussion of their content. The folk theories are primarily seen as part of the behaviour and interactions of nanotechnologists, rather than an occasion for debate about their substance. The observation of such behaviour then allows me to capture certain patterns, specifically the way 'insiders' try to realize new technology and see the world in terms of opportunities and obstacles. This allows me to understand the diagnosis of, and exaggerated concern about, public reactions to nanotechnology as a case of nanophobia-phobia. Such an understanding can inform interactions with nanotechnologists, and be taken up in communication in practical situations, increasing reflexivity.

Actually, we can speak of two levels of reflexivity. The first-order reflexivity of articulating folk theories which transcend immediate experience, and using them to orient action; and the second-order reflexivity of understanding interactions and behaviours including one's own – what I called sociological enlightenment. The clusters of folk theories that I will present and discuss tend to be limited to first-order reflexivity.

The wow-yuck pattern and learning from what happened with GMO

The first cluster of folk theories can be seen in almost canonical form in the presentation of Vicky Colvin (Director of the Center for Biological and Environmental Nanotechnology at Rice University, Texas) in hearings about the nanotechnology bill of the US Congress in April 2003. (A version of this quote returns in Kulinowski, 2004.)

> New developments in technology usually start out with strong public support, as the potential benefits to the economy, human health or quality of life are touted. At our centre we call this the 'wow index'. . . . At present, nanotechnology has a very high wow index. For the past decade, nanotechnologists have basked in the glow of positive public opinion. We've wowed the public with our ability to manipulate matter at the atomic level and with grand visions of how we might use this ability. All this 'good news' has created a growing perception among business and government leaders that nanotechnology is a powerful platform for 21st century technologies. The good news has given nanotechnology a strong start with extraordinary levels of focused government funding, which is starting to reap tangible benefits to society.
>
> However, every new technology brings with it a set of societal and ethical concerns that can rapidly turn 'wow' into 'yuck'. [example of genetic manipulation of crops, public backlash crippling the industry despite the lack of sound scientific data about possible harm] The failure of the industry to produce and share information with public stakeholders left it ill-equipped to respond to GMO detractors. This industry went, in essence, from 'wow' to 'yuck' to 'bankrupt'. *There is a powerful lesson here for nanotechnology.*
>
> [my italics]

There are two (interconnected) folk theories in Vicky Colvin's account. The first, visible in many other texts; is about genetically modified organisms (GMO), contrasting the backlash of the (late) 1980s and 1990s with the acceptance of its promise in the 1970s. The actual history is more complex, and deserves to be analysed as to its dynamics – including how present folk histories could emerge.[2] The important point here is that actors like Vicky Colvin do not make an attempt to check the history. The storyline about the so-called impasse she presents has become part of the repertoire about what can happen to new technologies.

The other folk theory is the wow-yuck pattern and its generality. Part of that folk theory is the painting of the public as fickle, linked to the assumption that there is always a period of 'wow' before the 'yuck' emerges. Even if the main example, GMO, may be more complex and thus not support the folk theory, the theory continues to function as a second-order expectation, of what to expect that will happen – the wow-to-yuck trajectory. The generality of this claim is important because it allows lessons to be drawn for the next case – here, nanotechnology – up to the attempt to avoid such a trajectory this time.

Interestingly, and indicative of the uses to which folk theories of general patterns can be put, is the curious dialectics in this line of thinking. The wow-to-yuck trajectory has to be painted as inevitable, so as to create a hearing for the message (the 'thesis'). But must then be reformulated as the result of mistakes, misguided behaviour, etc. (the 'antithesis'), so that one can define better behaviour and go for it to achieve the desired goal (the 'synthesis'). The better behaviour, in Colvin's presentation, is defined as having ELSA studies accompany funding and stimulation of nanotechnology.[3] She also emphasizes doing research on risks at an early stage (cf. Colvin, 2005). Education of the public, and some public engagement, is the other commonly proposed better behaviour. The assumption that this is where things went wrong with GMO is rarely checked. In other words, projections linked to an action perspective are more important than empirical support for the claims.

An important reason for the neglect of checks is the dominance of a particular definition of the situation – which is, in fact, a further bit of folk theory. On a number of places in Colvin's presentation, one sees a concentric problem definition of working on a new technology: focus on the new development and arrange further issue concentrically around it, to be taken up one after the other (Deuten et al., 1997). The folk theory behind this is that the technology is being developed anyhow, so the problem is to get it accepted. This is visible in the ease with which she talks about roadblocks (to nanotechnology commercialization), and how public acceptance is identified as one of these possible roadblocks. The 'insiders' in the centre do not have to understand the phenomena underlying these 'roadblocks'; they only have to overcome them.

For Colvin, this bit of folk theory was a rhetorical resource in her presentation, to better engage her Congressional audience. But the concentric problem definition, and the thinking in terms of roadblocks to desirable developments, is widespread, and has effects on actions. At first, technology developers and other insiders will not notice adverse signals, they continue to try and enrol others on the basis of promises as they see them. Only when there is resistance to their message will they start taking notice of the wider world.

A concrete example is the discussion of health and environmental risks of nanoparticles. When this issue was raised, and further highlighted by the ETC Group (2003), the immediate response was negation (in all senses of the word), and fury about the ETC proposal for a moratorium on nanoparticles. In a news feature article in *Nature*, it was noted that 'the debate is clearly gathering pace', while 'some researchers . . . feel that they don't need to join in the argument. "They don't really see what the hoop-la is about"' (Brumfiel, 2003, 247). The issue was relegated to an outer layer, as not relevant to what was important now.

Inputs from toxicologists and epidemiologists (and scientists like Colvin) introduced some moderation, but the gut reaction of nanotechnologists and other insiders remained. It was not legitimate to seriously discuss such risks, because that would only enlarge a possible roadblock. By the time the (so-called) Royal

Society Report appeared in July 2004 (Royal Society et al., 2004), with its message to be cautious with the introduction of nanoparticles in the environment because of the knowledge gaps about health and environmental impacts, it had become more difficult to just claim that nanoparticles were no cause for concern. The balance shifted, irreversibly, with the appearance of re-insurer Swiss Re's Report in August 2004. Discussing (and researching) risks of nanoparticles then became legitimate, even necessary. One aspect, played upon by ETC Group and Swiss Re alike, was 'size matters': if the small size is what gives nanoparticles their interesting properties, these same size-dependent properties can also create harm.

This third folk theory of nanotechnologists and other insiders, arranging the world concentrically around their effort to develop and introduce new technology, when recognized for what it is by analysts and commentators, then also explains the insiders' behaviour when encountering resistance. They will exclude, or at least relegate to outer layers, such resistance as illegitimate and thus not an occasion for further inquiry, until they are forced by further actions and reactions to do something about it.

The hype cycle as a widespread folk theory

The wow-yuck pattern, as Colvin called it, is about concerns, or at least second thoughts, after initial enthusiasm. Such a hype-disappointment cycle can also occur because of the almost inevitable failure of new technologies to live up to their early promise – which was exaggerated, or at least embellished, to get a hearing, and some support. Introduced by Gartner Group as the hype cycle for information and communication technologies, it has become a folk-theory *par excellence*, because widely recognized, used to draw out implications, and not an object of systematic research. The visualization provided by the Gartner Group is widely referred to and copied on websites. It shapes thinking about further developments and possible responses.

The storyline of the hype cycle starts with a technology trigger: a breakthrough, public demonstration, product launch, or other event generates significant press and industry interest. This leads to a peak of (more or less) inflated expectations: over enthusiasm, unrealistic expectations, and a flurry of well-publicized activity by technology leaders results in some successes, but more failures, because the technology is being pushed to its limits. A subsequent trough of disillusionment is pictured as inevitable: because the technology has not lived up to its inflated expectations, it rapidly becomes unfashionable, and the press either abandons the topic for the next hot thing or emphasizes its failure to meet expectations. A slope of gradual improvements – focussed experimentation and solid hard work by an increasingly diverse range of organizations – leads to a true understanding of the technology's applicability, risks, and benefits. Commercial off-the-shelf methodologies and tools become available to ease the development process and application integration. The real-world benefits are demonstrated and accepted. Tools and methodologies are increasingly stable as they enter their second and third generations.

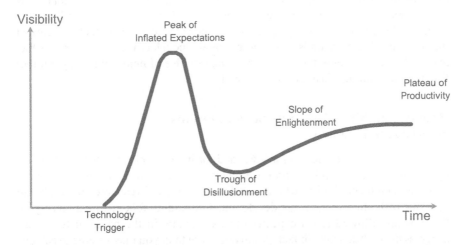

Figure 4.1 The hype-disappointment cycle as visible on websites[4]

This is how Gartner Group and their followers depict what happens. While it purports to indicate a recurrent pattern, i.e. an empirical finding, it is primarily a plausible storyline about how things go. The dialectics of patterns out there and actor strategies which were part of Colvin's presentation are visible here as well. If one 'knows' how things go, one can profit from that knowledge, just as venture capitalists profit from their insight in the rise and fall of promising new firms. For example get in while the hype is rising, and get out before the disappointment sets in. In fact, the Gartner Group has based part of its business model on the hype cycle by presenting and using it as a tool to offer strategic advice to companies. When you know, advised by the Gartner Group, where you are with your product(s) on the hype-disappointment cycle, you can define adequate strategies.

The graph and the term 'hype cycle' constitute a folk theory, not limited to the domain of information and communication technologies anymore. That it is a folk theory is clear from their emblematic character, for example leaving the vertical axis in the graph undefined; or when there is an indication (as 'visibility' or 'increased visibility & expectations' in the versions offered by the Gartner Group), it is not operationalized. Part of the force of a folk theory is that it is not fully articulated.

There are substantial issues linked to the phenomenon of hype and subsequent downswing, in information and communication technologies, in nanoscience and nanotechnologies, or wherever. In particular, about 'overshoot' ultimately damaging credibilities and reputations – while at the same time, some 'overshoot' seems unavoidable because early promises are necessary to reach audiences and markets. This introduces a further set of strategy considerations. Not about when and how to invest in something that is happening anyhow, but whether, as an initiator, to fuel the hype, or be more modest. Such considerations are part of resource mobilization strategies of scientists, when they have to indicate the promise of their projects and programmes. Nanoscience and nanotechnologies, at the moment, live on promises, so these strategy considerations are very visible, and lead to debates about how to present the promise of nanotechnology. Further folk theories and justifications are invoked.

Resource mobilization strategies and promises about nanotechnology

The mobilization of resources for R&D and for investments in new technological options necessarily builds on promises. This is a fact of life, also for nanoscientists and nanotechnologists who are mobilizing external resources to be able to continue with their research and development. Including folk justifications saying that you are allowed (even expected) to exaggerate, in the name of (eventual) progress through science. In fact, promising more than you may be able to achieve is not limited to nanoscience and nanotechnologies. It is generally accepted, and young researchers are enculturated in these practices. Their justification refers to an informal folk theory about the nature of science and how it should be supported by society.

These observations are an entrance point to discuss a second cluster of folk theories, centred around resource mobilization. The action perspective is not so much about learning from the past, as about scenarios of possible futures.

In general, strategies derived from the necessity of promising take two main forms: be modest to avoid hype and thus subsequent disappointment, or claim wonderful things and profit from the immediate advantages this brings. When genetic modification became serious as a technological option, around 1980, biotechnologists actually positioned themselves and their colleagues in these terms.[5] The same is now happening with nanoscience and nanotechnologies.

While the visions offered by nanoscientists and technologists, and other insiders, are generally optimistic, interestingly, different emphases are visible within the overall positive perspective. In the US, promises of a third industrial revolution and human enhancement abound, while in Europe there is less such talk. 'Nanotechnology is a new manufacturing technology able to make products smaller and stronger', that is how the spokespersons for the priority on nanotechnology in the EU's 6th Framework Programme present the promise of nanotechnology, in brochures and, interestingly, in a text in the *Cordis Newsletter* of June 2003, reporting on a meeting with ETC organized in the European Parliament on 11 June 2003. There, it was in the interest of the EU priority on nanotechnology (NMP, Nanomaterials, and Processes) to play down the revolutionary aspects which might raise concerns, and suggest nano was important, but nothing unusual, just a bit smaller, just a bit stronger.

Contrast this with how US Undersecretary of Commerce for Technology positions his message to the Swiss Re conference on nanotechnology of December 2004:

First, nanotechnology is coming, and it won't be stopped. . . . Second, given nanotechnology's extraordinary economic and social potential, it would be unethical, in my view, to attempt to halt scientific and technological progress in nanotechnology. Nanotechnology offers the potential for improving people's standard of living, healthcare and nutrition; reducing or even eliminating pollution through clean production technologies; repairing existing environmental damage; feeding the world's hungry; enabling the blind to see and the deaf to hear; eradicating diseases and offering protection against harmful bacteria and viruses; and even extending the length and the quality of life through the repair or replacement of failing organs. Given this fantastic potential, how can our attempt to harness nanotechnology's power at the earliest opportunity – to alleviate so many earthly ills – be anything other than ethical? Conversely, how can a choice to halt be anything other than unethical?

(Bond, 2005)

The biblical ring in some of the phrases is interesting in its own right. The quote can be taken as an indication of the US approach, because the setting, a small and closed conference, and talking to the convinced, was not one where Bond needed to make strong statements to reach his audience.

There is a dual folk theory here about symbolic and material resource mobilization and eventual outcomes. First, that hype is an integral part of the life of all new and emerging sciences and technologies, even if it can still take different forms. The modest or ambitious scenarios about future developments, when presented in public or semi-public spaces, will have implications for what nanoscience and technologies are expected to deliver. Diffuse 'requirements' (the term derives from procurement in the military sector) on nano R&D will emerge. And these may well set the scene for versions of the hype-disappointment cycle, because actual ongoing R&D has its own dynamics and is not tightly linked to these 'requirements'. Here, the other side of the dual folk theory about what scientists are allowed, even expected, to do to mobilize resources kicks in. The disjunction between what was promised, and what could actually be realized, was not only pardoned, and turned into further promises in the conclusions of articles and reports saying that further research was necessary but was glorified as exemplifying the open-ended and therefore cornucopian nature of science.

There are mechanisms to protect the micro-level of ongoing work from being checked against the claims made at higher levels. Up to the folk-theory saying that relabelling and 'shirking' (as principal-agent theory has it, cf. Morris, 2003) is allowed, even expected. But public scrutiny is increasing and may interfere with these traditional mechanisms. Even create a further backlash when the divergence between public claims and actual work is revealed.

The divergence between promises and actual work is not just an effect of the rhetorics of promising. R&D work will always have an element of bricolage, and that is one way in which R&D is productive, by not keeping to earlier envisaged plans of work. This general point takes on a special form in nanoscience and technologies, because the more interesting interventions at the nano-scale rely on induced self-organization, and thus have results that cannot be predicted. This has been emphasized by Jean-Pierre Dupuy, offering the image of nanoscientists and technologists as intentional sorcerer's apprentices (Dupuy, 2004; Dupuy and Grinbaum, 2005). Such an image contrasts with the promise of control over matter at the nano-scale that is a recurrent element in public statements about nanotechnology, and reflects a folk theory of what science is about, and should be about. This tension between an image of control, and a practice of trial-and-error bricolage, can be handled on the workfloor of the lab, but will create problems when products with particular properties have to be made reliably.[6] Thus, a structural basis for a hype-disappointment cycle, somewhat independent of the hype-embracing or modest strategies of actors.

The second cluster of folk theories is deeply engrained in the practices of science, and only articulated when these practices are interrogated. If outsiders start such interrogations, the insiders will often respond defensively. Another way to make such practices more reflexive is for analysts – or actors, for that matter – to create scenarios, for example for the overall development of nanotechnology. Combining the hype-disappointment pattern with different resource mobilization strategies, two main scenarios result:

1   Funding agencies, governments, and venture capitalists are willing to invest in promises, and nanotechnologists responding to these possibilities become intellectual venture capitalists, building on possibilities rather than what happens on the workfloor. The research and development topics offered are chosen for their potential to mobilize resources, rather than their likelihood of actual delivery. This is a self-reinforcing dynamic – until something goes wrong, and the bubble bursts.
2   There is less interest from venture capitalists, and this is fuelled by the strong stand of the modest nanotechnologists who position the hype-embracers as 'street vendors'. Links with production actors rather than decision actors are pursued. Improved materials and drug delivery are prominent, roadmapping is taken seriously.

To get a hearing for such a scenario exercise (and the implied articulation of folk theories), they can be linked to actor's concerns. In my contribution to the preparation of the Royal Society Report (2004), I argued that these two scenarios actually reflect what is happening in the US and Europe, respectively. Thus, if (when) the US bubble bursts, Europe will be in a better competitive position – an important actor's concern.

This example indicates one particular way for analysts to respond to actors' folk theories and actions building on them: show what might happen given their explicit and *de facto* strategies. To make a difference, feedback to nanoscientists and technologists is essential. In my own case, being an integral part of a consortium of nanoscientists I have privileged access – but still have to work to earn my hearing.[7]

In the scenarios, and in the cluster of folk theories that they draw upon, the public remains in the wings. As if there were a protected space where nanoscientists and technologists can just interact with other insiders, with sponsors of nano R&D and industrialists of various ilk. There is fierce competition within this protected space, but also a shared concern to avoid outside interference. At the same time, it is clear that such outside interference will occur.

There appear to be two main routes for insiders to interact with the outside (to their protected space). One route is to identify further stakeholders and spokespersons for civil society, and create dedicated interactions. This has happened with GMO technology (cf. Doubleday, 2004), and is visible for nanotechnology in attempts to engage with critical groups. The other route is to engage with publics more generally, in nano-dialogues, in nano-juries, in scientific cafés. It is a way to mobilize symbolic resources which hopefully ensure support for further development of nanotechnology. These are ongoing projects, set up from the bottom (as with scientific cafés) or incited by top-down funding (the EU supported nano-dialogues).

Various folk theories are at play in these activities, with deficit models for public understanding (members of publics as empty vessels to be filled by explanations of what nanotechnology is really about) still holding sway. When the boundary of the protected space is transgressed, it is on the terms of the insiders. The deficit model of public understanding of science includes a folk theory about insiders as the real experts. I will come back to this folk theory, in particular how it can be positioned and used to improve interactions between insiders and outsiders to the nano-world. But I will first draw attention to the behaviour of nano-insiders with respect to actual and potential public reactions, as they interpret them.

Nanophobia-phobia

In the dominant folk theory of scientists and technologists about the public and its reactions, members of the public are seen as empty vessels, to be filled with understanding of science to avoid emotional reactions running riot. The use of the singular, 'the' public, is another indication of projections, homogenizing what is heterogeneous and dynamic, and thus allowing the view of the public as fickle to continue. When discussing Colvin's folk theories, this view of the public was already visible.

There is a third cluster of folk theories here. They may be evolving. Analysts like Brian Wynne have played a key role in the UK and elsewhere in delegitimating the deficit model so that public engagement has become more legitimate (cf. Wilsdon and Willis, 2004). Whether this is a matter of insiders responding to credibility pressures, or an actual shift in the folk theories is less clear. The rational-emotional divide continues to be a natural way (and a resource) for scientists and technologists to position the public.

Consider the results of a workshop organized by the Nanotechnology Working Group of the UK Royal Society and Royal Academy of Engineering to find out about the views of scientists and engineers about nanotechnology (specific areas and general). What is said under the sub-heading 'Social and Ethical Issues' (in the report in the background documents of Royal Society 2004) reflects the traditional perspective:

> A general point was made relating to the public's perception and media reporting of nanotechnology. It was felt that in order to conduct a rational debate, a realistic projection of the potential impacts (positive and negative) of nanotechnology must be communicated to the public. It was felt that hyped up reports from some scientists or writers have only served to confuse the public's perception of nanotechnology. This has caused confusion between what is fact and what is fiction, and may create unjustified fears. Key messages that the group felt should be put in the public domain were that nanoparticles are not new – very small particles have always been around, and that nanotechnology is multidisciplinary and an enabling technology rather than a new discipline in its own right. The best solution to these public fears was seen as better public understanding at a scientific level through the provision of better science information in school and university undergraduate curricula.

There are lots of projections without any empirical support in this quote. Insiders permit themselves to expostulate in this way because of a sense of certainty about what is the case – an indication of a folk theory being around.

A further effect of this folk theory about the public is that nano-insiders start to expect that concerns will be voiced (in their view, because of limited understanding of what nanoscience and technology is really about). Interestingly, even if there is little or no public concern (yet), some nanotechnologists see reference to, and indications of, such concerns everywhere, even in innocent phrases as 'nanotechnology: little things with big consequences'. This particular example emerged in the discussions of a working party of the Royal Netherlands Academy of Sciences, established to respond to a request of the Minister of Education, Culture, and Science to advise about potential of, and issues around, nanoscience and technology, including the possible need to look into societal aspects of nanotechnology. The working party, chaired by a toxicologist, had six members: five nanoscientists and an STS scholar (Arie Rip). One of the nanoscientists strongly objected to the use of the term 'consequences' in the title, which he was certain would suggest to the public that there was something to be concerned about. Eventually, another, more neutral, title was chosen 'How big can small be?' (KNAW, 2004).[8]

Thus, there is not only an exaggerated interpretation of concerns as voiced, seen as an indication of fear, even phobia of the new technology. Such concerns and fears are also projected on the public when there is no occasion. In other words, a folk theory about the public and its reactions, which resurfaces again and again – and in spite of the fact that the limited data available indicate appreciation of the new nano-ventures rather than concern.[9] Thus, the concern of nanoscientists and technologists about public concerns (painted as a phobia about nano) drives their views, rather than actual data about public views. For some of them, like the nanoscientist in the working party of the Netherlands Academy, their concern has become a phobia in its own right, seeing fear where none might exist.

The concern about possible public's concern is getting a life of its own. A small but revealing finding (ETC, 2004, 7) is that on the Internet, the discussion of concerns about Grey Goo (and Crichton's novel *Prey*) is not conducted by the public, but by nanotech actors and other insiders and commentators concerned about possible public reactions – even if these are still absent. There are also first results of opinion surveys, showing that reading Crichton's novel makes people more interested and positive about nanotech, rather than more negative (Cobb and Macoubrie, 2004; Hanssen and Van Est, 2004). Insiders find this hard to believe.[10]

In other words, there is a general presumption that publics are passive and susceptible to fearful interpretations (here, after reading a science fiction novel). Specifically, scientists and technologists (and other promoters of nanotechnology) are prone to project nano-phobia, and this projection can become a phobia in itself, a nanophobia-phobia.

This reaction is not limited to nano, and may well be a general feature of views and behaviours of insiders. There was widespread concern about ill effects of chemistry in the 1970s and 1980s, and this led chemists – who love chemistry – to think in terms of a chemo-phobia. As a knowledgeable commentator (a science journalist) noted at the time, the chemists were so concerned about chemo-phobia that they saw it everywhere, even when there was little or no cause. They were building up a chemophobia-phobia.

In addition to indications of chemophobia-phobia which I could observe in my moving about in the world of chemists, we did a small study to check whether and how it occurred (Rip and Slot, 1982). We identified a number of attitudinal statements about chemistry, and asked respondents to indicate to what extent they agreed or disagreed. One set of respondents was a (small) sample of the general public. Another set of respondents were chemists. For the latter set, we also asked them to fill in the responses they thought the general public would give to these statements. The results showed that chemists were more positive about chemistry than the general public, but also that the general public was not very negative about chemistry, and definitely less negative than the chemists thought they would be. The largest differences between the public's responses and the chemists' projection of what they would be occurred for statements about trust in the profession of chemists. This supports the diagnosis of chemophobia-phobia: insecurity about trust leads chemists to see distrust everywhere, even when there is no reason to do so.

As with the chemists, there is insecurity with nanoscientists and technologists about the trust in their professional role, which fuels their phobia. The strong promises about nanotechnology compensate for this insecurity, so the phobia may remain limited in scope. In diagnosing nanophobia-phobia I have switched from describing a folk theory of nanotechnologists about reactions of publics, with action implications, to a diagnosis of the behaviour of nanoscientists and nanotechnologists, at least some of them. Nanophobia-phobia may be an extreme case. But it is a useful entrance point to develop a broader diagnosis of the situation of nano-insiders. This diagnosis will help us to understand the structure of the folk theories of nanotechnologists, and why these theories are so tenacious.

Enactors and their enactment cycles

Let me rephrase the thrust of the previous section in less dramatic terms. Nano-scientists and technologists want to realize what they think is important, starting with their own ongoing work and the financial and moral support for it. Thus, they become concerned about every signal that indicates concern about, let alone resistance, to their mission. They will see roadblocks, also when these are not there. And sometimes they create communication problems by their actions responding to actual or projected roadblocks.

Drawing on a seminal study by Garud and Ahlstrom (1997), on TA in the health sector, these views and actions can be understood as linked to the position of the scientists and technologists as insiders. Garud & Ahlstrom show that there are different socio-cognitive perspectives linked to two basic positions:

A   *Insiders* who try to realize or 'enact' new technology, construct scenarios of progress to be made, and identify obstacles that must be overcome. Thus, they work in 'enactment cycles' (p. 410). Such 'enactment cycles' emphasize the positive aspects of the new option, and work through an illusion of control.

B   *Outsiders* who are faced with this option, but see also other options (up to the null option of not going for any innovation), pursue 'selection cycles' in which they can compare the option with alternatives. Sometimes, as in the case of an agency like the US Food & Drug Administration, they are in a position of explicit external control.

Garud and Ahlstrom's terminology of 'insiders' and 'outsiders' can be improved upon. This way of characterizing actors assumes a boundary between an inside and an outside. In fact, the key point of their own analysis is the difference in socio-cognitive position and style of activities. So one had better speak of 'enactors' and 'comparative selectors' instead. The more suggestive terminology of 'insiders' and 'outsiders' can only be used if there are good reasons to distinguish between an inside and an outside. Then it will also become clear that there are *immediate* outsiders like the US Food and Drug Administration who have a mandate to be selective, and *distant* outsiders like spokespersons for societal groups and the general public who voice support or concern and exert selection pressure in this way.[11]

With Garud and Ahlstrom, I emphasize that there is a structural difference between the action perspectives and related views for the two positions. Messages from the 'enactors' to other actors will be framed in their enactment perspective, and not take the comparative-selection perspective into account. This explains us why scientists & technologists and other enactors of new technology assume that explaining the nature of the new technology as they try to realize it is enough to persuade other actors and the general public to accept it – and then be disappointed and frustrated that this does not work, up to comments from their side that the public must be irrational and emotional.

Garud and Ahlstrom suggest that 'bridging events' occur where the cycles linked to the two perspectives interact, sometimes productively. An attempt to communicate with the public can be a bridging event. Viewed in this way, communication is not a one-way process, where enactors decide what to tell the public. The enactors might be interrogated, and/or anticipate the comparative-selection perspective held by their audience.

Obviously, this background analysis, a proto-theory of an analyst, can (and should) be developed much further. For this article, it is sufficient to recognize the two positions and the different perspectives connected to them as structuring interactions, as well as constituting actors' folk theories.

In conclusion

The nanotech world is full of folk theories, so much is clear. I have discussed two clusters of folk theories. One cluster is about wow-yuck patterns, the need to learn from the GM impasse, and the concentric view seeing the world in terms of opportunities and roadblocks. The other cluster is about the mandate of scientists working for progress, which allows them to exaggerate promises when mobilizing resources, but makes them vulnerable to hype-disappointment cycles. There is first-order reflexivity, with actors recognizing this situation and articulating different strategies to respond to the situation. I have also identified folk theories of nanotechnologists (and of scientists and technologists generally) about the public and their reactions, and used the phenomenon of nanophobia-phobia to highlight one behavioural consequence. This was a stepping stone to outline a social science theory about insiders and their perspectives and behaviour, as enactors of new (science and) technology, with a self-defined mandate to do so.

Some of the folk theories are arguably wrong, as when they are based on projections of what happened with GMO, and how one should avoid a similar sorry fate for nanotech. But whether one or the other folk theory is wrong or not is less important for my overall argument than that they frame views and are connected to an action perspective. In other words, they are part of somewhat forceful repertoires. There are different repertoires at work, depending on the type of actors involved, and these repertoires evolve over time.

Enactors, among themselves, evolve a *repertoire* of promises and other expectations, and *strategies* how to position the future of nanotechnology and their own role in it. The repertoire need not be homogenous, and there are choices, for example between a modest strategy and one of embracing hype. Positioning the future of nanotechnology evolves in response to changing circumstances. One example is how 'regular' nanotechnologists try to exclude Eric Drexler and his followers from the nano-world (see also Berube and Shipman, 2004). This particular struggle, disavowing one of the prophets of nanotechnology, is linked to a change in the resource situation. By 2000, when the US National Nanotechnology

Initiative became operational, nanoscientists, and technologists in the US, but also elsewhere, were assured of funding – and did not need Drexler's visions anymore to mobilize support.

Part of the repertoires, and partially structuring them, are the various folk theories. They are about how science works, how technology develops, how public(s) behave, and can thus be compared with sociological and psychological theories and insights. The link of folk theories with action implies that there are further action-oriented folk theories, as of nanoscientists and technologists, and other insiders, about how to deal with outsiders. Or about patterns that tend to occur but can be overcome by the 'right' action, for example with the wow-yuck pattern.

That there are hype-disappointment cycles, and that these can be induced by over-optimistic promises of enactors and their take-up in decisions of sponsors (up to venture capitalists), is now accepted and taken into account. The separate question is *how* to take this into account. There are two modes of reflexive reaction of actors. One is about strategic choices, whether to be modest, or embrace the hype. The other is about the dialectics: as soon as a pattern is identified, this induces further reflection on how to escape it, or undermine it, or influence it, so as to further one's own interests and perceptions.

Social scientists, and in particular STS scholars, can play a role in enhancing reflexivity. While folk theories are conservative, almost by definition, they do evolve. To a large extent, this occurs from the outside in, under pressure of changing circumstances including threat of loss of credibility.

Part of such interactions will be with STS scholars, directly or indirectly. An example of the latter is the effect of a publication like *See-through Science* (Wilsdon and Willis, 2004), e.g. when taken up in editorials in an authoritative journal like *Nature* (2004). Thus, STS scholars will not just have an intellectual role. They will also act, or be positioned to act, as spokespersons for changing circumstances. And, given the nature of folk theories in the nanoscience and technology world, also be seen as spokespersons for 'the outside'. For nanoscientists and technologists, inviting social scientists in means taking society into account. There is a further action-oriented folk theory here, assumed by scientists and technologists: the role of social scientists is that of a 'lubricant' between science/technology and society (Newby 1992), and their contribution is positioned and evaluated as serving a better – in the sense of smoother – introduction of new science and technology in society.

At the same time, a new role for social scientists is emerging: being invited in implies getting a better hearing, and interacting at a much earlier stage of developments than usual. This new role is taken up systematically in the approaches of constructive technology assessment (Schot and Rip, 1997) and real-time technology assessment (Guston and Sarewitz, 2002). And is reflected upon in Fortun (2005, 157, 159, 161), when he calls on social scientists and humanist scholars to be 'friends with the scientists'. For nanoscience and technology, this is very visible. In the Netherlands, social scientists participate in the nanoscience and technology consortium NanoNed.[12] In other countries, one sees interactions in the way US centres for nanotechnology in society have developed their research programmes, and joint sponsorship of activities, as with the nano jury in the UK, an initiative of a newspaper (The Guardian), the Nanoscience Research Centre in Cambridge, and an NGO (Greenpeace UK).[13]

These recent developments indicate that co-evolution of nanoscience/ technology and society is becoming more reflexive, and that sociological enlightenment may play a role. The co-evolution continues to be heterogeneous and conflictual, inevitably so. And learning occurs, but often only because of contestation and threats. Folk theories are a fact of life, in general and in the world of nanoscience and technology in particular. They also offer an entrance point for further interaction with social scientists because claims are made about what is the case and will be the case (at least at the level of second-order expectations), and about good reasons for doing some things and avoiding others. These can be taken up in structured debate and other bridging events.

Notes

1 Science, Technology, and Society (STS) scholars can, for example, criticize the linear model of innovation. They position themselves as knowing better. On what grounds? Basically, because they have moved about more widely than the actors themselves. But they have their own folk theories as well, for example the QWERTY keyboard as an emblem of path dependency, and the overpasses on Long Island as an emblem of political shaping of technology. For a discussion of the latter as a folk theory, see Joerges (1999a), Woolgar and Cooper (1999), and Joerges (1999b).

2 See for such analysis Grove-White et al. (1997) and Wynne (2001). One aspect of the complexity is the role of companies rather than the public, for example the role of Monsanto's becoming assertive about its strategy at the same time it had chosen pesticide resistance rather than disease resistance. A reconstruction of Monsanto's role is overdue (cf. Gorman et al., 2004, 55–56; Doubleday, 2004). In addition, one should note that for genetic engineering, critical debate – the yuck in Colvin's folk theory – was present from the beginning (Krimsky, 1982).

3 'In contrast [to GMO], the Human Genome Project provides a good model for how an emerging technology can defuse potential controversy by addressing it in the public sphere. . . . They wisely welcomed and actively encouraged the debate from the outset by setting aside 5 percent of the annual budget for a program to define and address the ethical, legal and other societal implications of the project.

I sincerely hope that we can learn from this example. . . . In effect, early research into unintended consequences redirects the wow-to-yuck trajectory. . . . We seek to avoid the path traveled by the GMO industry by encouraging the industry to answer the tough questions about societal and environmental impacts while it is still developing'.

Colvin continues, 'We need partners in this endeavor. Based on the recent National Research Council report and our own experience, there is little money and interest in the societal, ethical and environmental impact of nanotechnology, despite the rhetoric. Your help here is essential. . . . If I had to guess, I would estimate that of the nearly one billion dollars slated to go to nanotechnology this year not even one percent is directed specifically towards studying the societal, ethical and environmental impact of nanotechnology. A tangible symbol of your commitment to this kind of research would be to set a target research funding for the area; the 3 to 5 percent rule used by the Department of Energy in the Human Genome Project would be a good starting point'. The Nanotechnology Act of November 2003 actually reflects this sentiment (*21st Century*

Nanotechnology R&D Act of 2003, 108th US Congress, First Session, S.189, Section 5 b and c), even if there is no dedicated ELSA funding specified.

4 Versions of the hype-cycle were presented by Gartner Group since at least 1999, see Fenn (1999).

5 With biochemists and molecular biologists making strong claims, and microbiologists and chemical engineers, closer to actual biotechnology, being more modest. Compare Rip and Nederhof 1984.

6 *The Economist* (March 2003) identified two big challenges for nanotechnology. The first challenge is the carbon-silicon interface, i.e. the interaction between living entities and electronic devices. The second is the unpredictable behaviour of nanoscale objects. This means that engineers will not know how to make nanomachines until they actually start building them. There will be trial and error, and success will be a matter of luck rather than design.

7 The Dutch NanoNed consortium, with government funding and contributions from the participating universities, research institutes, and a big firm (Philips), includes a technology assessment and societal aspects component. Background studies and interactive (constructive) technology assessment workshops articulate societal aspects, and ensure some feedback to nano-actors. See www.nanoned.nl.

8 Another struggle was about the inclusion of a detailed refusal (up to picking on typos) of Crichton's novel *Prey*, so as to show that it was unscientific – which was assumed sufficient to lay to rest the fears of the public thought to be occasioned by reading *Prey*. This particular struggle was won: there was no such page in the final report.

9 Survey and focus group data which show this include the background study by BMRB International for the Royal Society 2004 Report, and Cobb and Macoubrie (2004). I have also used an unpublished overview of results of focus group studies and further indications of attitudes and views of publics (Hanssen and Van Est, 2004).

10 Interestingly, this latter finding was used in a little quiz at the beginning of a hearing about nanotechnology organized by the Rathenau Institute for the Dutch Parliamentary Committee on Technology, 13 October 2004 (see Van Est and Van Keulen (2004) about this meeting). The 60 or so participants had to respond to slides with multiple-choice questions about nanotechnology, by raising their hands to vote. Only participants with the right answer could go on and respond to the next slide, until a single winner remained. After three slides, there were still 50 or so active. Then came the *Prey* slide, with the three possibilities of people becoming more negative, kept their (neutral) view, or became more positive. Only ten participants remained for the next round. This may be statistically significant but is definitely anecdotally significant.

11 In terms of conceptualization of positions, there is partial similarity with analysis in terms of (tolerance to) the uncertainty trough, as introduced by MacKenzie (1990) and developed further in Rip (2001).

12 Website www.nanoned.nl, TA/ELSA.

13 Website www.nanojury.org

Bibliography

Atran, S. (1998), Folk Biology and the Anthropology of Science: Cognitive Universals and Cultural Particulars, *Behavioral & Brain Sciences*, 21(4), pp. 547–570.

Berube, D., and Shipman, J.D. (2004), Denialism: Drexler vs. Roco, *IEEE Technology and Society Magazine*, 23(4), pp. 22–26.

Bond, P.J. (2005), Responsible Nanotechnology Development, in *Swiss Re Centre for Global Dialogue*, pp. 7–8. Zürich: Swiss Re.

Brumfiel, G. (2003, July), A Little Knowledge . . ., *Nature*, 424(17), pp. 246–248.

Cobb, M.D., and Macoubrie, J. (2004), Public Perceptions About Nanotechnology: Risks, Benefits and Trust, *Journal of Nanoparticle Research*, 6, pp. 395–405.

Colvin, V.L. (2003), *Testimony of Dr. Vicki L. Colvin, Director Center for Biological and Environmental Nanotechnology (CBEN) and Associate Professor of Chemistry Rice University, Houston, Texas Before the U.S. House of Representatives Committee on Science in Regard to 'Nanotechnology Research and Development Act of 2003' April 9, 2003*, www.house.gov/science/hearings/full03/apr09/colvin.htm

Colvin, V.L. (2005), Could Engineered Nanoparticles Affect Our Environment?, in *Swiss Re Centre for Global Dialogue*, pp. 19–20. Zürich: Swiss Re.

Crichton, M. (2002), *Prey*, London: Harper Collins.

Deuten, J.J., Rip, A., and Jelsma, J. (1997), Societal Embedment and Product Creation Management, *Technology Analysis & Strategic Management*, 9(2), pp. 219–236.

Doubleday, R. (2004), *Political Innovation: Corporate Engagements in Controversy Over Genetically Modified Foods* (Unpublished Ph.D. thesis, University College London).

Douglas, M.L. (ed.) (1982), *Essays in the Sociology of Perception*, London etc: Routledge & Kegan Paul.

Douglas, M.L., and Atran, S. (eds.) (1999), *Folkbiology*, Cambridge, MA: MIT Press.

Dupuy, J.-P. (2004), *Complexity and Uncertainty: A Prudential Approach to Nanotechnology*. A contribution to the High-Level Expert Group 'Foresighting the New Technology Wave'. Brussels: European Commission, March 2004. Printed in Nanoforum electronic publication *Benefits, Risks, Ethical, Legal and Social Aspects of Nanotechnology*, www.nanoforum.org

Dupuy, J.-P., and Grinbaum, A. (2005), Living With Uncertainty: Toward the Ongoing Normative Assessment of Nanotechnology, *Techne*, 8(2), pp. 4–25.

ETC (2003), *No Small Matter II: The Case for a Global Moratorium – Size Matters!* Occasional Paper Series 7(1).

ETC (2004), *May/June 2004 Communiqué* (Issue # 85).

Fenn, J. (1999) *When to Leap on the Hype Cycle: Research Note*, Stamford, CT: Gartner Group, www.cata.ca/files/PDF/Resource_Centres/hightech/reports/indepstudies/Whentoleaponthehypecycle.pdf

Fortun, M. (2005, August), For an Ethics of Promising, *New Genetics and Society*, 24(2), pp. 157–173.

Garud, R., and Ahlstrom, D. (1997), Technology Assessment: A Socio-Cognitive Perspective, *Journal of Engineering and Technology Management*, 14, pp. 25–48.

Gorman, M.E., Groves, J.F., and Catalano, R.K. (2004), Societal Dimensions of Nanotechnology, *IEEE Technology and Society Magazine*, 23(4), pp. 55–62.

Grove-White, R., Macnaghten, P., Mayer, S., and Wynne, B. (1997), *Uncertain World: Genetically Modified Organisms, Food and Public Attitudes in Britain*, Lancaster: Lancaster University in association with Unilever.

Guston, D.H., and Sarewitz, D. (2002), Real-Time Technology Assessment, *Technology in Society*, 24, pp. 93–109.

Hanssen, L., and Van Est, R. (2004), *De dubbele boodschap van nanotechnologie. Een onderzoek naar opkomende publiekspercepties*, Den Haag: Rathenau Instituut.

Heintz, C., Pouscoulous, N., and Taraborelli, D. (eds.) (2007), Special Issue of *European Review of Philosophy* (Vol. 8, 2007) on Folk Theories. Website, www.erp-review.org, visited 27 March 2006.

Joerges, B. (1999a), Do Politics Have Artefacts? *Social Studies of Science*, 29(3), pp. 411–431.

Joerges, B. (1999b), Scams Cannot Be Busted: Reply to Woolgar and Cooper, *Social Studies of Science*, 29(3), pp. 450–457.

KNAW (2004), *Hoe groot kan klein zijn? Enkele kantekeningen bij onderzoek op nanome-terschaal en mogelijke gevolgen van nanotechnologie*, Amsterdam: KNAW (Koninklijke Nederlandse Akademie van Wetenschappen). English translation included.

Krimsky, S. (1982), *Genetic Alchemy: The Social History of the Recombinant DNA Controversy*, Cambridge, MA: MIT Press.

Kulinowski, K.M. (2004), Nanotechnology: From 'Wow' to 'Yuck'? *Bulletin of Science, Technology and Society*, 24(1), pp. 13–20.

Lakoff, G. (1987), *Women, Fire, and Dangerous Things: What Categories Reveal About the Mind*. Chicago: Chicago University Press.

MacKenzie, D. (1990), *Inventing Accuracy: A Historical Sociology of Nuclear Missile Guidance*, Cambridge, MA: MIT Press.

Morris, N. (2003), Academic Researchers as 'Agent' of Science Policy, *Science and Public Policy*, 30(5), pp. 359–370.

Nature (2004, October), Going public, 431(21), p. 883. Editorial.

Newby, H. (1992), One Society, One Wissenschaft: A 21st Century Vision, *Science and Public Policy*, 19(1), pp. 47–54.

Rip, A. (2001), Contributions From Social Studies of Science and Constructive Technology Assessment, in A. Stirling (ed.), *On Science and Precaution in the Management of Technological Risk: Volume II. Case Studies*, Sevilla: Institute for Prospective Technology Studies (European Commission Joint Research Centre), pp. 94–122.

Rip, A., and Nederhof, A. (1986), Between Dirigism and Laisser Faire: Effects of Implementing the Science Policy Priority for Biotechnology in the Netherlands, *Research Policy*, 15, pp. 253–268.

Rip, A., and Slot, W. (1982), Beelden van Chemie, *Chemisch Magazine* (okt), pp. 600–602.

Rip, A., and Talma, S. (1998), Antagonistic Patterns and New Technologies, in C. Disco and B.J.R. Van der Meulen (eds.), *Getting New Technologies Together*, Berlin: Walter de Gruyter, pp. 285–306.

Royal Society and Royal Academy of Engineering (2004), *Nanoscience and Nanotechnologies: Opportunities and Uncertainties*. RS Policy Document 19/04. London: Royal Society.

Schot, J., and Rip, A. (1997), The Past and Future of Constructive Technology Assessment, *Technological Forecasting and Social Change*, 54, pp. 251–268.

Sumner, W.G. (1907), *Folkways: A Study of Mores, Manners, Customs and Morals*, reprint published by Mineola, NY: Dover Publications, .

Swidler, A. (1986), Culture in Action: Symbols and Strategies, *American Sociological Review*, 51, pp. 273–286.

Swiss Re (2004), *Nanotechnology: Small Matter, Many Unknowns*. Risk Perception Series, Zürich: Swiss Reinsurance Company.

Swiss Re Centre for Global Dialogue (2005), *Nanotechnology: "Small Size – Large Impact?"*. Risk Dialogue Series, Zürich: Swiss Reinsurance Company.

Van Est, R., and Van Keulen, I. (2004), "Small Technology: Big Consequences": Building up the Dutch Debate on Nanotechnology From the Bottom, *Technikfolgenabschätzung. Theorie und Praxis*, 13(3), pp. 72–79.

Wilsdon, J., and Willis, R. (2004), *See-Through Science: Why Public Engagement Needs to Move Upstream*, London: Demos.

Woolgar, S., and Cooper, G. (1999), Do Artefacts Have Ambivalence? Moses' Bridges, Winner's Bridges and Other Urban Legends in S&TS, *Social Studies of Science*, 29(3), pp. 433–449.

Wynne, B. (2001), Creating Public Alienation: Expert Cultures of Risk and Ethics on GMOs. *Science as Culture*, 10(4), pp. 445–481.

5 Emerging *de facto* agendas surrounding nanotechnology

Two cases full of contingencies, lock-outs, and lock-ins

Arie Rip and Marloes Van Amerom

Published as

Arie Rip and Marloes van Amerom, Emerging *de facto* Agendas Around Nanotechnology: Two Cases full of Contingencies, Lock-outs, and Lock-ins, in Mario Kaiser, Monika Kurath, Sabine Maasen, Christoph Rehmann-Sutter (eds), *Governing Future Technologies. Nanotechnology and the Rise of an Assessment Regime*. Dordrecht etc: Springer, 2010, pp. 131–155.

Introduction and conceptualization

In a number of ways, the development of nanoscience and nanotechnologies is more reflexive than was the case for earlier new and emerging sciences and technologies. One indication is the common reference to the so-called impasse around (green) biotechnology, and how to avoid a similar impasse (for an example, see Colvin 2003; for an analysis in terms of folk theories, Rip 2006c). Related to this is the willingness to invite public engagement, if only as a precautionary measure. There is also reference to the importance of 'responsible' development of nanoscience and nanotechnologies, e.g. in European Commission documents and in recent initiatives for voluntary codes. Clearly, there is now space for reflection and deliberation.

Does this imply that deliberations will play a constitutive role in the formation of nanotechnologies? The role and effect of deliberations will always be predicated on the emergence of openings for deliberation in the ongoing co-evolution of nanotechnology and society, and the links with ongoing societal agenda-building. Thus, we need to understand the dynamics of co-evolution, the patterns that emerge, and in particular, which overall agendas become *de facto* dominant.

We present two case studies of these dynamics. The first case is about how Eric Drexler, once positioned as a founding father of nanotechnology, became excluded from mainstream nanoscience and nanotechnology. The fate of Drexler and his view is linked to the discussion of 'molecular manufacturing' and, in relation to this, the possibility of a 'Grey Goo' scenario.

The second case is about the emergence of potential hazards of nanotechnology, in particular of nanoparticles, as a legitimate concern. By 2006, there were concrete actions and reactions, ranging from regulatory agencies exploring what to do about nanoparticles, to some firms becoming reluctant to work with nanoparticles.

In both cases, there is discussion and debate, but not necessarily deliberation in the strong sense. On the other hand, some learning occurs in such controversies (Rip 1986), somewhat independently of the emergence of spaces for explicit deliberation. To take this into account, we use the metaphor of an evolving socio-technical landscape. 'Landscape' indicates the backdrop against which actions and interactions are played out, which enables and constrains and thus shapes what happens. The landscape evolves, partly because of stabilization of certain agendas and arenas, actor constellations, and patterns in interactions. The key point is that 'landscape' is not just a passive backdrop against which humans play out their affairs. It is itself constructed, and part of the 'play' is to construct elements of the backdrop.

This conceptualization is like Giddens' notion of structuration, but now at the meso-level and with attention to actual dynamics. These dynamics include the build-up of socio-technical infrastructures and how they recede into the landscape.[1] Just as gradients in a landscape (say, hills, and valleys) shape the movements of people and other 'mobiles' that traverse the landscape, a socio-technical landscape shapes action and perception. It can be seen as a tangible story, with routing devices to guide the 'reader' without determining the reader's movements. Some of these routing devices have evolved naturally, and almost all of them are outcomes, at a collective level, of a variety of actor strategies, designs, and interventions, which to some extent (and after some time) are unintended by any of the actors. The landscape is a *dispositif*, just as much as the more explicitly socio-technical *dispositifs* studied by Foucault and others.[2]

A visualization of such a landscape is Sahal's (1985) diagram indicating trajectories of evolving innovations (Figure 5.1).[3]

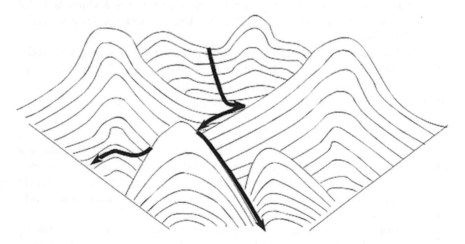

Figure 5.1 Topography of socio-technical evolution (Sahal 1985: 79)

While the diagram conveys a clear message about technology dynamics, for our purposes it can be misleading. The landscape is already full of paths and forks, and it is their (agonistic) interaction (for example a battle about an industry standard) which shapes outcomes. For our broader use of the 'landscape' metaphor, we must add further dynamics, in particular *de facto* agenda building. We are interested in the contour lines in Sahal's diagram, indicating gradients that enable and constrain, and how these came about, rather than in one or another particular path. We will map the evolving 'contour lines' (broadly speaking) in the following case studies, which will allow us to understand the what is happening, as well as to consider possible future paths enabled and constrained by the evolving landscape.

Societal *de facto* agenda-building interests us, rather than the traditional focus in agenda-building analysis on one single arena and what happens inside. Societal agenda-building is a multi-arena process, and does not have a clear authority deciding on the agenda.[4] Kingdon (1984) is a good starting point for such analysis, because of his discussion of policy entrepreneurs and their skills, their networks, and how they can act on policy windows, openings, or opportunities to forge a new or change the existing agenda. An additional factor is how issues can become linked so that new alliances emerge (such as around radioactive waste burial around 1970, cf. de la Bruhèze 1992). Such (always partial) entanglements are a general phenomenon,[5] and they can become locked in and lead to path dependencies – which are themselves an example of *de facto* agenda setting and stabilization (cf. Rip et al. 2007).

Existing agendas, dominant discourses, and actor constellations are a backdrop to ongoing processes, e.g. emerging actor constellations around an issue, which promote stabilization of certain agendas – which thus changes the landscape. Still, 'windows of opportunity' will occur, albeit fewer than before (cf. Stirling 2005). One circumstance reducing flexibility is how arenas stabilize by excluding actors that are no longer considered legitimate spokesmen. This is particularly evident in our first case, and it is reinforced by actors using (and relying on) stereotypical characterizations, in this case, of the Drexlerian view. Similarly, after 2006, risks of nanoparticles were generally expected to exist, while the uncertainties and lack of evidence underneath this characterization were black-boxed.

In our second case, the entanglement of actions, reactions, and emerging discourses and constellations is particularly evident. The health, environmental, and safety (HES) aspects of nanoparticles are now high on the agenda in the 'nano world'. They can thus be seen as a priority, and their implementation a subject of inquiry. But these directions to go emerged from earlier entanglements, which included ongoing work on risks and debates on regulation. In other words, what can now be positioned as implementation of an agenda on risk started before such an agenda was in place. To understand such processes, where (so-called) implementation happens prior to goal-setting,[6] one has to reconstruct the processes' dynamics, rather than follow the linear histories that are produced by actors to support their current efforts to push for more and better risk research (e.g. Maynard et al. 2006).

This brief discussion of societal *de facto* agenda setting, together with our earlier (and technology-dynamics-inspired) consideration of evolving landscapes, allows us to visualize our approach as a multilevel characterization of interactions and entanglements leading to patterns and agendas that shape further action, but can also open up and shift. Figure 5.2 is of course a simplification, but it conveys a message about the importance of interactions, and especially interactions at the mid-level. All of these interactions add up to an evolving landscape with an overlapping patchwork of contours rather than one definite set.

In this visualization, the focus is on the meso-level and interactions with micro- and macro-levels. Larger 'framework conditions' that are also part of the landscape are not indicated. Similarly, when we reconstruct our two cases and offer visualizations of their dynamics (in Figures 5.3 and 5.4 below), we will not explicitly include such conditions as changing regulatory cultures (acceptance of precautionary measures, shift towards 'soft law'), the role of disciplinary cultures (differences between chemistry and engineering, for example), and specifics of national cultures and structures.

To trace these multilevel landscape dynamics empirically will not be simple. One can isolate a specific question and apply standard social science operationalization and data gathering. The challenge, however, is to trace interactions and how these add up to up to a composite picture. In a sense, this is

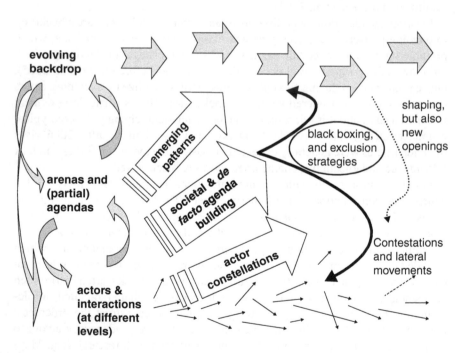

Figure 5.2 Multilevel landscape dynamics

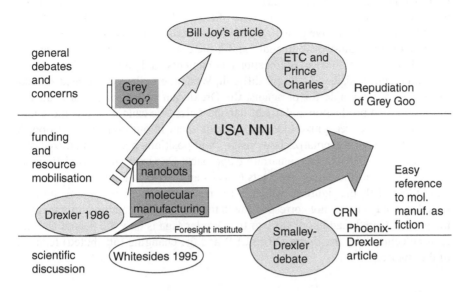

Figure 5.3 The evolving landscape of the Drexler vision

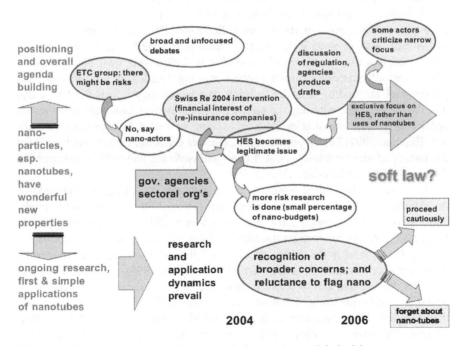

Figure 5.4 Evolving paths in the 'landscape' of nanotubes and their risk

writing contemporary history. Thus, we have also relied on blogs and newsletters, which offer gossip; however, in the case of Howard Lovy and Tim Harper (TNT Weekly), it is authoritative gossip. We have also used our own experience in moving about in the nano-world. In addition, for the case of risks of nanoparticles, we created our own database of publications, documents, and commentaries.

Contemporary history writing is difficult. What one writes can be read as 'taking sides', for example for or against the Drexlerian vision, or for or against a moratorium on production and use of nanoparticles. We cannot escape these tensions, but we can recognize them and include them in our conceptualization. A way to do this is to thematize how visions and positions become stabilized and black-boxed. This is, in fact, a narrative approach. One example is to distinguish between Drexler, the person who acts (speaks and writes) in concrete situations, and DREXLER, the figure to which is referred and on which features are projected that may or may not correspond with the actual behaviour and utterances of the person. One thread in our story of the Drexler saga is how the person Drexler is often eclipsed by the figure DREXLER and the changing (attributed) features of that figure.

The Drexler saga

We were struck by the ease with which Richard Jones, nanoscientist, and commentator, could say (and be accepted saying it), 'Drexler, of course, is the name that can't be spoken in polite society'.[7] 'Polite society', of course, is the society of the mainstream nano-world, and there may be a tinge of irony in the way Richard Jones phrased his comment. But it does indicate how Drexler has become a figure: as Howard Lovy phrased it in correspondence with Eric Drexler, quoted on his blog, 'Like it or not, the name of "Drexler" is no longer your own' (Lovy 2003b). This is the outcome of a complex process in which 'leaders in both industry and government are finding it easier to bring nanotechnology out of the fringe and into the mainstream, whetting the public's appetite with rudimentary commercial applications' if they 'cast aside Drexler's vision, as well as his warnings' (Berube and Shipman 2004: 24). In popular texts, Drexler can still be positioned as one of the fathers of nanotechnology, but if nano-actors do so, they will be relegated to the camp of the 'Drexlerians'.[8]

There are, by now, quite a number of articles and book chapters that analyse the Drexlerian vision and its ambivalent fate (Kaiser 2006; Bennett and Sarewitz 2006; Berube and Shipman 2004; Bensaude-Vincent 2006; Milburn 2004, 2008; Selin 2007). As we reconstruct the history, there are two different but connected strands. One strand is the debate on the feasibility of molecular manufacturing, which became increasingly antagonistic from the late 1990s onwards. The other strand is the rise and fall of concern about the Grey Goo scenario, linked to the possibility self-replicating 'nanobots'. The debate peaked in 2003 and 2004, and then receded. These two strands may be linked to the exclusion of Eric Drexler from the main nanoscience and technology arena, from 2000 onward. After

Drexler's claims about the principle possibility of molecular manufacturing (or assembly),[9] and the possibility of the Earth being turned into Grey Goo by the replicators needed for such assembly, had become topics of contention in the early 2000s, they now appear to be topics of non-contention. Thus, the stakes involved must have settled. In any case, now that funding for nanoscience and nanotechnology is assured, there is no longer a need for a Drexlerian prophet of nanotechnology. This is, at least, Drexler's own understanding of what happened (Drexler 2004).

In this section, we will first note how Drexler himself linked molecular assembly and a Grey Goo scenario in his *Engines of Creation* (1986). In the 1990s, the discussion of Drexler's speculative vision was constructive.[10] A first shift occurred around 2000, with contestation of the vision and positioning of the Grey Goo scenario as a concern. Both stabilized by 2004, together with the generally accepted exclusion of the Drexlerian vision from mainstream nanoscience and nanotechnology. To understand this stabilization, we will also trace the discussion in and around the US National Nanotechnology Initiative (NNI).

In Drexler's *Engines of Creation* (1986), he describes 'molecular assemblers' as devices capable of building products from the atom up, thus with absolute precision and without pollution. However, in order to do so (i.e. produce amounts that are visible and useful macroscopically), there must be lots of molecular assembling going on. So, these assemblers must also reproduce themselves. Assuming that the first assembler could make a copy of itself in 1,000 seconds,

> The two replicators then build two more in the next thousand seconds, the four build another four, and the eight build another eight. At the end of ten hours, there are not thirty-six new replicators, but over 68 billion. In less than a day, they would weigh a ton; in less than two days, they would outweigh the Earth.
>
> (Drexler 1986: 172–173)

And so they would consume this Earth in the process. In other words, if the replicators are instructed to copy themselves, and there is no built-in stopping point, they will eat up everything and turn the Earth into Grey Goo (i.e. a jumble of replicators). While the notion of Grey Goo was discussed, for example in the sci. nanotechnology newsgroup, and referred to occasionally in the media, it only became part of an emerging societal debate on possible drawbacks and risks of nanotechnology after Sun Microsystems' founder Bill Joy made a plea in 2000 to constrain the development of converging technologies. Grey Goo became an image referred to in newspapers worldwide to imagine and discuss 'nanotechnology' dangers (Anderson et al. 2005). Another input was the publication of Michael Crichton's 2002 science fiction novel *Prey*, which drew further attention to the notion of Grey Goo, although it was about out-of-control swarms of biological organisms created through nanotechnology (*Los Angeles Times 2002*; ETC Group 2004: 7). Nanoscientists all over the world were concerned about public and political reactions.[11]

Nanotechnology risk stakeholders who were demanding precautionary approaches to nanotechnology referred to, and imagined nanotechnology dangers in terms of, Grey Goo (Munich Re 2002; Arnall 2003; ETC Group 2003b). When it seemed that the UK's Prince Charles was concerned about nanotechnology because of Grey Goo fears (in April 2003), there was a new peak of media coverage on Grey Goo, inside and outside the UK (Feder 2003; Thurs and Hilgartner 2005).

Whether there was indeed public fear of a Grey Goo scenario or not (and there are indications that it was more a phobia from the nano-actors about the public's fears than actual public nano-phobia; Rip 2006c),[12] NST promoters were concerned that 'popular fear of Grey Goo would be a harbinger of a general backlash against nanotechnology' (Hilgartner and Lewenstein 2005), and started to publicly deny the possibility of Grey Goo and attack and ridicule those who believed in it. Events in the UK are illustrative. Following the *Mail on Sunday*'s (27 April 2003) assertion that Prince Charles was concerned about nanotechnology in relation to fears over Grey Goo, Buckyball co-discoverer Sir Harry Kroto accused the prince of 'a complete disconnection from reality'. Lord Sainsbury, Minister for Science and Innovation, publicly denied the feasibility of the Grey Goo scenario. Chairman of the House of Commons Science and Technology Committee Ian Gibson reproached Prince Charles merely for mentioning Grey Goo: 'We shouldn't be associated with scare stories – science fiction about grey goos . . . [because] when a prince speaks, people will listen' (Oliver 2003a, 2003b).

This part of the story has been told before (though not always in this specific way). The big debate about the feasibility of molecular assembly, in particular debates between Nobel Prize winner Richard Smalley and Drexler, 2001–2003, appears to have been conducted independently of the Grey Goo scenario.[13] Smalley's arguments about 'fat and sticky fingers' hinge on the fact that molecular assemblers are conceived as mechanical (cf. below). Drexler's counter-argument has been to refer to the assembling that goes on all the time inside living cells: it's natural, so it must be possible in principle. In this arena, there is only passing reference to the Grey Goo scenario. It is only later, when the Grey Goo scenario is picked up by critical organizations such as the ETC Group (the Canadian-based Action Group on Erosion, Technology, and Concentration), and ascribed to visible actors like Prince Charles, that the link becomes active.

In the scientific realm, the discussion was about the feasibility of Drexler's ideas on molecular manufacturing. After relative inattention in the late 1980s and early 1990s, except in a circle of enthusiasts mostly at the Foresight Institute established in 1986, the discussion was taken up in a somewhat appreciative manner (Whitesides 1998, cf. also 2001). After open interactions at Foresight Institute conferences in 1995 and 1997, Smalley started to question Drexler's visions of molecular assemblers, and then actively attacked them in an article in *Scientific American* (2001). Smalley maintained that manipulator fingers on the 'hypothetical' self-replicating nanobot would be 'too fat' to pick up and place individual atoms with precision and 'too sticky' to let them go after having picked them up (Smalley 2001: 68). Drexler's response was that his visions of molecular manufacturing never envisaged nanobots, making the 'fat fingers' or 'sticky fingers' problem irrelevant. The debate continued, and attracted wide attention when the protagonists had their say in *Chemical & Engineering News* on 1 December 2003 (Baum 2003). While to some extent inconclusive, the debate was seen by US nanotechnology business and government actors, as well as many scientists keen to distance themselves from what they could now call science fiction, as a victory for Smalley (cf. Lovy 2003a). Drexler's ideas could be declared to be unfeasible, to the frustration of the Foresight Institute and other 'Drexlerian' actors like the Center for Responsible Nanotechnology (CRN), who continued to appeal to the more speculatively minded.

Around the same time, in October and November 2003, a provision of the US 21st Century Nanotechnology Research and Development Bill relating to molecular assembly was at stake in the struggle over how to position Drexler's vision.[14] The initial (House) version of the bill included a provision (written by California Rep. Brad Sherman, a Drexler supporter) calling for a study to evaluate the technical merits of 'molecular manufacturing', and, if possible, prepare a timeline and a research agenda.[15] The Senate was less keen on such a provision, but the phrasing created a flurry of protests in the nanoscience community, and possibly pressure on members of Congress from representatives of the US NanoBusiness Alliance and other NST promoters (Regis 2004).[16] In any case, the bill's final version now referred to 'molecular self-assembly' and asked for 'a one-time study to determine the technical feasibility of molecular self-assembly for the manufacture of materials and devices at the molecular scale'. As commentators noted, self-assembly is a known process (and therefore 'innocent'), but the key question is the interpretation of the subsequent clause on manufacturing at the molecular scale. Mark Modzelewski of the NanoBusiness Alliance (when interviewed by *US News & World Report*) said, 'It is possible that some aspects of "molecular manufacturing" might be investigated, but knowing the parties influencing the study, I doubt it. There was no interest in the legitimate scientific community – and ultimately Congress – for playing with Drexler's futuristic sci-fi notions' (quoted in Lovy 2003a, cf. also TNT Weekly 2003). Henceforth, Smalley's arguments became a key reference for the dismissal of Drexler's visions on molecular assembly, but somewhat independent from the debate about Grey Goo.

Thus, there were three arenas of debate and strategizing: developments around science and funding for nanotechnology, the nanotechnology risk debate in society, and the feasibility of molecular manufacturing. They are connected to one another, and references to Drexler as well as the activities of Drexlerians like the Foresight Institute and CRN are part of the connection. In other words, there are links, but they are not linear. Delegitimizing Grey Goo scenarios by arguing that molecular manufacturing is science fiction is one possible strategy, and it can be linked to attempts to exclude Drexler's visions, and thus Drexler, from 'polite society'. But other strategies and linkages are possible as well. Over time, though, a particular constellation of attributions and positioning can become dominant, and such a dominant constellation will then have a definite set of strategies and linkages to justify its actions.

By 2004, such a dominant constellation was emerging. While the 21st Century Nanotechnology Research and Development Act of December 2003 allowed for government-funded studies on the possibility of Grey Goo (Fisher and Mahajan 2006), there were no steps in this direction. Emerging government-funded research programmes into the ethical and social implications of nanotechnology in the US did not include the possible detrimental effects of molecular manufacturing as a research focus. Similarly in the UK, a report from the Royal Society commissioned by the UK government simply could state that the possibility of Grey Goo was not worth researching because Drexler's molecular manufacturing ideas had proved to be mere fiction (Royal Society/Royal Academy of Engineering 2004: 109).

The dismissal of a Grey Goo scenario also occurred at the side of nanotechnology risk-alerters: these identified possible risks, but these were not linked to Grey Goo scenarios. For example the ETC Group still listed Grey Goo as a possible NST concern in 2003 (ETC 2003a) and, in the midst of the publicity surrounding Prince Charles' alleged Grey Goo concerns, criticized the dismissal of Grey Goo by NST promoters (ETC 2003c). However, in July 2004, in another communiqué, the activist group denied having ever spoken about Grey Goo, instead blaming NST promoters for having brought the concept of Grey Goo into circulation (ETC Group 2004). In the same month, Prince Charles publicly distanced himself from the Grey Goo scenario. More than a year after the public reference to his alleged Grey Goo fears in British newspapers, he denied, in an article in the *Sunday Independent*, having ever believed in the possibility of Grey Goo (HRH The Prince of Wales 2004). The statements may well reflect how ETC Group and Prince Charles used the notion of Grey Goo in the past as a means to draw attention to possible risks of nanotechnology. Clearly, there was no need to do so anymore in 2004, and such a reference might become counterproductive, isolating them from mainstream opinion. This is a tactical move, but it reinforces mainstream views, and will be seen and used as such.

The 2004 article written jointly by CRN director Chris Phoenix and Eric Drexler can be seen as an attempt to re-enter the nanotechnology arena as legitimate players by distancing themselves from Grey Goo, or to 'Un-Goo' and align visibly with more legitimate concerns. There are substantial arguments in the article as well, but the article was received as a concession of defeat and a conversion to mainstream thinking. Phoenix and Drexler's argument is that, thanks to new technological developments, nano-manufacturing no longer needs autonomous self-replicating nanomachines. Military use (or abuse) of nano-manufacturing appears a more immediate threat (Phoenix and Drexler 2004: 869). Thus, there should be more attention to the security aspects of nanotechnology. Anti-Drexler nanotechnology promoters, however, portray the article as Drexler finally admitting that the prospect of Grey Goo had been a mere 'fantasy' (for example Institute of Physics 2004), which adds to his lack of credibility. Journalists accepted and copied this interpretation (see for example Rincon 2004; Sample 2004; Sherriff 2004; *The Scotsman 2004*). Thus, instead of overcoming Drexler's exclusion from the mainstream nanotechnology arenas, the article is used to continue his exclusion.

We visualize the dynamics in Figure 5.3 below (note that the overall backdrop to these dynamics is not indicated).

Looking back, it is clear that there are two turning points in the developments: 1999–2000 and 2003–2004. In fact, these are turning points in the overall development of nanotechnology. For 1999–2000, Bennett and Sarewitz (2006: 312–313) show a steep rise in media interest. The 2000 NNI in the US ensured and coordinated substantial government funding for nanotechnology research and, more generally, helped 'the development of the burgeoning nanotechnology industry' (Guzman et al. 2006: 1401). The NNI indicated acceptance of nanotechnology, signalling its establishment as a field of research and setting a model for other countries to follow. Having obtained funding and political legitimacy, the construction of nanotechnology as an interdisciplinary but independent scientific field could begin in earnest. One further element of the situation was the competition for available funding by actors often holding very different views of nanotechnology. This led to implicit and explicit contestation over whose views were legitimate and feasible, and therefore deserving funding.

Boundaries were being drawn as to what comprised and did not comprise nanotechnology, as well as who were 'legitimate' nanotechnology players and who were not (Kaiser 2006). All this is a common feature of emerging professions (Abbott 1988: 60). In the course of this process, Drexler's ideas on molecular manufacturing, which had been one of the guiding visions for nanotechnology (Robinson et al. 2007; Wood et al. 2007: 3), were dismissed as being too 'far out' for the profession, and thus had to be redefined as fictional. As Milburn (2004: 118) phrases it, 'Nanotechnology managed to secure its professional future by combining fantastic speculation with concerted attacks on science fiction'. (The latter is a rhetorical ploy to show that nanotechnology is a real science.)[17] 'Regular nanoscientists' and nanotechnology business players phrased their promises in terms of what has been called 'near-term nanotechnology' (Peterson 2004: 10): scanning tunnelling microscopy and atomic force microscopy, nanotubes, supramolecular chemistry, and new ways of etching and constructing thin layers, as these techniques were perceived as being able to produce results relatively fast and enable the usage of nanotechnology for commercial purposes.

The connection of Drexler's visions to the Grey Goo debate became a further argument to exclude Drexler and the Drexlerians. For example Modzelewski, spokesman of the Nanotechnology Business Alliance, accused Drexler of being 'irresponsible' by thus endangering the development of nanotechnology and its enormous benefit for humankind. Smalley also used Drexler's concern about Grey Goo as a stick to hit him, accusing him of needlessly 'scaring our children' with 'scary stories' (Smalley 2003: 42). Drexler's association with Grey Goo was used as a moral justification for his 'demonization' by the science and business community.[18]

Boundaries are not made once and for all, however. Ongoing developments on what are now called molecular machines, including natural and artificial molecular motors, are hailed as 'a significant step towards future nanomachines and devices' (Browne and Feringa 2006). Thus, speculation about molecular manufacturing continues and is actually taken seriously, as long as the linkage with Drexler's vision is not emphasized. And even that is not problematic anymore: the UK Engineering and Physical Science Research Council has granted £2.5 million to 'invent a nanomachine that can build materials molecule by molecule' (Van Noorden 2007). The approach is not Drexlerian (it relies on scanning probes which induce self-assembly), but one of the scientists involved is willing to say,

> If it works, it will redefine nanotechnology as it should have been . . . referring to concepts promoted in the 1980s by U.S. engineer Eric Drexler, who suggested that nanotechnology would create tiny machines dubbed 'assemblers' that could drag atoms and molecules around to make copies of themselves, or other useful devices.
>
> (Van Noorden 2007)

It is significant that Drexler's ideas have to be explained: the present generation of nanoscientists is assumed not to be aware of them.

Allis (2007), a Drexlerian interested in concrete experiments, cashes in on these recent developments:

> You've got single Si atom manipulation, Feringa's optical motors, Tour's got his nanocar. Those things aren't dimer deposition to build diamondoid gears [a Drexlerian option], but they're far more 'mechanical' than chemists were thinking 30 years ago, and they certainly hint at all the potential we have for fundamental control over matter.

It is doubtful whether these developments will lead to a rehabilitation of Drexler's vision; the exclusion pattern has become institutionalized. This shows that our visualization in Figure 5.3 should include ongoing nanoscience research and the expectations that are voiced about the research. From 2007 forward, we might see a revival of molecular manufacturing (under other labels), at least as laboratory curiosities.

Health, environmental, and safety aspects of nanoparticles

The emergence and recent broad acceptance of the acronyms ELSA (Ethical, Legal, Social Aspects) and HES (Health, Environmental, Safety)[19] in discourse on and governance of nanotechnology research indicates emerging stabilization of HES issues. The force of HES is itself the outcome of what could be labelled an emerging and stabilized path, at the meso-/macro-levels. It is instructive to reconstruct its emergence, including the contingent elements.

Around 2000, the broad promises of nanoscience were pushed – up to 'shaping the world atom by atom' (National Science and Technology Council 1999) – and some concerns about the development of nanotechnology were starting, partly in response to the big promises voiced by and about the new US NNI. The other generally visible issue was Bill Joy's 2000 warning about future technologies, including reference to the Grey Goo scenario (cf. preceding section). Two NSF-sponsored workshops on opportunities and societal implications collected a variety of essays, some of them referring to concerns about side effects of the development of nanotechnologies, including a 'nano-divide' (between the global North and the South) and military usage and a new arms race (Roco and Bainbridge 2001; Roco and Tomellini 2003; see also Roco 2003).

On the environmental front, a Center for Biological and Environmental Nanotechnology (CBEN) was established at Rice University, and its scientists reported in March 2002 to the Environmental Protection Agency that engineered nanomaterials might accumulate in the human body, as well as potentially cause environmental degradation.[20] One indication of how little articulated that discussion was at the time is the January 2002 report by one of the big re-insurance companies, Munich Re: it raised concerns about risks in general terms, and there was very little response to its message – in contrast to the worldwide response a similar report, by Swiss Re, two years later produced.[21]

A first focus emerged when the ETC Group issued a communiqué, *No Small Matter*, in July 2002, in which its general concern about new technologies was applied to nanotechnology. The fact that action was proposed (specifically, to stop making nanomaterials until we know more about environmental impacts, and have this debated at the level of the UN) and the responses to that proposal (cf. *Small Times* 2002) was the beginning. The ETC Group followed this up with a report, *The Big Down*, in January 2003, which called for a moratorium on the commercial production of new nanomaterials (2003a). The immediate response was negation: denial that there could be risks, and denial that the ETC Group should be listened to. There was also fury about the ETC proposal for a moratorium on nanoparticles (*Small Times* 2003a). While in a news feature article in *Nature*, it was noted, 'The debate is clearly gathering pace', while 'some researchers . . . feel that they don't need to join in the argument. "They don't really see what the hoopla is about"' (Brumfiel 2003: 247).

In the first half of 2003, the attention level for nano-risk issues increased together with the number of actors entering the arena. In April, the Woodrow Wilson International Center started its Project on Emerging Nanotechnologies, including a study on risk and regulation of nanoparticles. Also in April, US congressional hearings on a new nanotechnology bill were the occasion for various actors to propound their message, including Vicky Colvin referring to lessons from the history of genetically modified organisms and the need, therefore, to anticipate and do research on potential risks (Colvin 2003). Later that same month, the news broke about Prince Charles' concerns (cf. preceding section), which contributed to the UK government, in June, commissioning and funding the Royal Society and Royal Academy of Engineering to have an in-depth look at these issues. The UK Institute of Nanotechnology, together with the Green Party members of the European Parliament and the ETC Group, organized a meeting in Brussels in June, to which staff of the European Commission felt it had to respond, informally.

Thus, there was an atmosphere of contestation. Illustrative is how nanotechnology promoters were prepared to dismiss, out of hand, the report commissioned by Greenpeace UK that came out in July that year (Arnall 2003). The report showed interest in the possibilities of nanotechnology, and did not call for a moratorium (it does note that a moratorium on engineered nanoparticles might prove to be necessary, should industry not invest more in nano-risk research). US promoters of nanotechnology called the report misleading propaganda and said it was 'too early to have these kinds of discussions' (Rob Atkinson, Progressive Policy Institute). Modzelewski of the NanoBusiness Alliance called it 'industrial terrorism', while New York State Rep. Sherwood Boehlert (House Committee on Science and Technology) said, '[On] issues like this there will always be people on the extreme' (*Small Times* 2003b).

After the first-round denial of risks, and in spite of the ongoing occasional strong condemnation of concerns over these risks, 2003 can still be marked as the beginning of a next phase: at this point, possible risks of engineered nanomaterials were recognized, but these could be looked into while development continued (no real harm was expected). It was not legitimate to seriously discuss action implications of such risks, because that would mean a roadblock to further development. Inputs from toxicologists and epidemiologists (and scientists like Colvin) introduced some moderation.

The legitimacy of concerns about risk increased also because of first research results presented by Oberdörster to a 2004 meeting of the American Chemical Society which were widely reported as to their implications, as well as criticized as to methodology (see Oberdörster et al. 2005) produced response. Toxicologists defined research needs, and government actors (including in particular the European Commission) started to move to explore the issue. The ETC Group published an overview of relevant research in one of its occasional papers (ETC 2003b). The scientific arena was becoming active.

The balance shifted, irreversibly, with the appearance of re-insurer Swiss Re's report in May 2004, with its strong linking of risks of asbestos and risks of nanotubes, and nanoparticles in general (see also Menon 2004). Discussing (and researching) the risks of nanoparticles then became fully legitimate. One paradox, played upon by the ETC Group and Swiss Re alike, was that 'size matters': if their small size is what gives nanoparticles their interesting properties, these same size-dependent properties can also create harm.

A specific 'risk hierarchy' emerged, with most actors, at least officially, agreeing that the nanoparticle issue would be the most important and urgent risk on which to concentrate, and the notion of Grey Goo being (re)framed as a 'fictional' concern and a form of science fiction. While nanotechnology promoters now followed 'prudent' scientists and nanotechnology risk alerters by recognizing the importance of, and engaging with, the nanoparticle risk issue, the other side had moved as well: nanotechnology risk alerters and reporters followed nanotechnology promoters in their rejection of Grey Goo as an issue (cf. preceding section).

By the time the Royal Society/Royal Academy of Engineering report appeared in July 2004, its message could not be dismissed: the introduction of nanoparticles into the environment requires caution because of the knowledge gaps about health and environmental impacts. Various nano-promoters did continue to critically evaluate ongoing research, and on that basis argue that there was still little cause for concern.

One further indication of the emerging closure was the establishment and composition of the International Council for Nanotechnology (ICON) in October 2004. Initiated by Rice University as a network to 'assess, communicate, and reduce' HES risks of nanoparticles, it was able to include not only other research institutions, governmental agencies, and non-governmental organizations (NGOs), but also representatives of the NanoBusiness Alliance. Corporate actors such as L'Oréal, DuPont, Procter and Gamble, and Unilever were founding sponsors. Clearly, there was sufficient common ground to have multi-stakeholder collaboration at this point.

The same growing agreement on the framing of risk issues (though not necessarily on what to do) enabled and was then reinforced by broadly inclusive meetings. One example was a major workshop organized by Swiss Re in December 2004. Also, working groups were set up, particularly by OECD. The UK government responded to the Royal Society/Royal Academy of Engineering Report and globally endorsed it. Initiatives to risk management became legitimate and consultants saw a business opportunity (e.g. Lux Research 2005; see also Nordan and Holman 2005). After earlier explorations, regulatory initiatives could now be considered. And nanotechnology promoters turned around, the most striking example being US House Committee on Science's Chairman Boehlert (who had earlier condemned Greenpeace UK for putting risk on the agenda), who called for more funding into nano-risk research, noting, 'This is the time to act, before we cause problems. This is the time to act, when there is a consensus among government, industry, and environmentalists' (PhysicsOrg.com 2005). Knowledgeable commentator Tim Harper (Cientifica) saw a 'safety bandwagon' emerge (Harper 2006).

After stabilization of such a common ground – with a strong focus on risk, particularly of engineered nanoparticles – government agencies, NGOs, and companies started to engage with practicalities. For government agencies accepting a precautionary approach (Rip 2006a), there was the challenge of regulating without knowing what exactly to regulate. One approach then was to start with existing regulatory frameworks and apply them, perhaps while modifying them. The other approach was to address the uncertainty as such, for example by introducing a voluntary reporting scheme, as UK Department for Environment, Food, and Rural Affairs (DEFRA) did in 2006. This would be an example of 'soft law' (Trubek et al. 2006).

For companies, the advent of regulation resolves part of the uncertainties. In a world where HES risks may now be expected to occur, there is also the uncertainty about consumer reactions. If something untoward happens somewhere under the umbrella term 'nanotechnology', that negative event will have repercussions for other products, even if these are (presumably) safe.[22] After the earlier marketing tactic of using the 'nano' label for products, firms started to become more careful and delete 'nano' from the labels of their products, or stop their line of nano-containing products altogether. Those who continued proceeded cautiously, and were willing to consider voluntary codes of conduct, so as to show good practice.

In Figure 5.4, we visualize part of the dynamics (they are too complex to depict in one diagram), exemplified by nanotubes, and add some further features such as criticism of the present focus on risk.

One interesting phenomenon is how arenas overlap and how actor roles become hybrid. Government actors with regulatory responsibility (especially when they are pro-active) attend meetings and generally take part in a variety of arenas where informal societal agendas are built. Similarly, industrial actors mingle with other kinds of actors, especially if a somewhat neutral space is provided. An interesting example is the meeting organized by Swiss Re and the International Risk Governance Council in Zürich in July 2006.[23] The occasion was the publication of a (hopefully authoritative) report on risk governance of nanotechnology, authored by Ortwin Renn (a risk and public deliberation scholar) and Michael Roco (of the US NNI) (see also Renn and Roco 2006a). Governmental and industry actors from across the world attended, as well as NGOs, scientists, and scholars studying nanotechnology in society. Dedicated workshops and mingling in the corridors allowed interaction and recognition of positions of other actors (and thus some learning).

The traditional distinction between formal agenda-building by authoritative (policy) actors and informal societal *de facto* agenda-building becomes blurred. According to Shibuya (1996), for a risk issue to rise on the formal agenda, it needs to be taken up in both formal and informal agenda-building processes. However, the articulation and prioritization processes are not separate. This is why we needed the concept of an evolving landscape to map the processes. It also shows how earlier and ongoing actions and interactions about risk and governance can be repositioned, after stabilization of the agenda, as activities to implement newly articulated goals, which in this case are responsible innovation and risk governance.

In this evolving landscape, two paths are visible: one is the focus of concerns on risk, and in particular risks of nanoparticles; the other is the tendency to opt for soft law in the interaction between governments (and their agencies and advisers) and firms (and their associations, sectoral or otherwise, and alliances). While these two paths shape most of what is happening now, there are also other paths. One is the criticism of the narrow focus on risk and risk assessment, and not just by NGOs like Greenpeace. The other is the involvement of NGOs in the soft-law alliances, for example in the proposal by DuPont Company and the US non-profit Environmental Defense Fund for a voluntary risk-assessment framework that can be adopted by oversight agencies worldwide.[24]

There is now also increasing reference to 'responsible innovation' in government documents (particularly of the European Commission) and some industry statements. While this may invite nano-promoters to consider broader issues, and allow other actors to raise questions about directions of development, responsible innovation is presently operationalized as transparency and some public engagement. And in the case of industry, also as a responsibility for safe handling of nano-production and nano-products.[25] The recent September 2007 initiative toward a 'Responsible Nanotechnologies Code' is led by the UK Royal Society, an NGO (Insight Investment), the Nanotechnology Industries Association, and supported by a network organized by the UK Department of Trade and Industry,[26] envisages a broader approach, but it is not clear if and how it will be taken up. (For an overview of the present situation, see Kearnes and Rip 2009.)

In conclusion

Our two cases show different patterns. The Drexler saga had an auspicious start, and Drexler's ideas inspired many people, including some of the later critics of the Drexlerian vision. But then the concept became weakened because its feasibility was a matter of in-principle argumentation, too 'long term' or 'speculative' or even 'science fiction', depending on the position one wanted to take. The link with the Grey Goo scenario did not help either. Boundary work to exclude Drexler, or better, DREXLER, the stereotyped carrier of the Drexlerian vision, started in earnest after funding for nanoscience and nanotechnology was assured in 2000, and the exclusion was complete by late 2003. As Regis (2004) phrases it, 'Drexler found himself marginalized in the very field he had inspired', while the spectre of DREXLER overshadowed his attempts to remain a player. Clearly, by now, molecular manufacturing in the Drexlerian sense is a path not taken.

Where the Drexlerian agenda has collapsed for all practical purposes,[27] the concern-about-nanotechnology agenda has become stronger, and is operationalized and implemented in its more focussed, or reduced, version of attention to risks, in particular risks of nanoparticles. The reversal occurred in 2004, from an open-ended situation of broad concerns and denial and contestation, to acceptance of risks of nanotechnology as a legitimate issue, and a situation in which government agencies as well as other actors should take concrete steps to prevent harm. By 2006, there was no way back, and as outlined in the preceding section, two paths had emerged in the overall landscape which shaped most of the activities.

Thus, in both cases, the entanglements led to *de facto* irreversibilities: no return for DREXLER; a continuation of the focus on risk and risk governance. The emergence of such irreversibilities is not a linear process, even if in retrospect one can tell a story of actions, interactions, and events leading up to the present situation. Also, there are contingencies, for example the effect of Prince Charles' interest in nanotechnology in spring of 2003, which triggered concrete actions by other actors. However, the effect of what might be seen as a contingent action or event is predicated on, and to some extent shaped by, the contours of the landscape at the time. Another example is how Swiss Re's May 2004 report on risks of nanoparticles was a singular intervention and had effects as such. But if it had been written and published two years earlier, it would not have had these effects because the landscape was not yet amenable, as is clear from the very different reception of the München Re report of 2002.

Before inquiring into spaces for explicit deliberation in these developments, it is important to note that learning is occurring at the societal level, or at least at some collective level. One of us has called this 'repertoire learning' (Rip 1986), in the sense that a better (more articulated) repertoire has developed that is available to actors to use (and misuse) and that will shape what are acceptable positions and actions. While struggles about definitions and directions linked to interest-driven strategies and overall agenda-building dynamics will not disappear, these struggles will be conducted in terms of the better articulated repertoire and thus be more productive. This productivity may well be predicated on following by-then stabilized patterns, as Swierstra and Rip (2007) have shown for moral arguments about newly emerging technologies such as genetics and nanotechnology. Therefore, openings and lateral action might be important to keep the repertoire evolving and adapting (cf. Rip 2006b).

There might be occasions where decontextualized deliberation, as in an (idealized) agora (cf. Nowotny et al. 2001), occurs, often orchestrated and supported by professionals, as in recent public engagement exercises about new technology. On these occasions, the possibilities and outcomes are determined more by the spaces that open up and allow for some deliberation, than by the deliberative processes and arguments as such. For example, prudent nano-promoters will be interested in longer-term issues (for example to avoid the impasse of green bio-tech), and thus prepared to entertain broader interactions and deliberations. This enables mutual learning. But learning is an effort, and actors will thus invest in learning only when they are forced to do so, to ensure their survival and/or to meet contestation. This applies to industrialists just as much as it applies to critical NGOs. In other words, while deliberative processes can assume that actors are able to learn and shift positions without much constraint, the real-world dynamics are full of emerging irreversibilities and stabilizing gradients. Shifts are always part of struggles.

The next step is then how to take repertoires and their dynamics, including the nature of the spaces, into account. We positioned and analysed our two cases not just to understand what happens while outcomes are not determined linearly by intentions and actions of one or more actors. We showed how actions and interactions at the collective level (overlapping arenas, *de facto* agendas) are important in determining outcomes. This offers tools (at least by example) to reconstruct and diagnose what is happening, which can then be fed back to various actors. We have developed this approach one step further by creating scenarios of further developments, based on nonlinearity and complexity (see Rip and Te Kulve 2008 for a first overview). In workshops with heterogeneous participants, as we have organized, such scenarios are starting points to probe each other's worlds, rather than working towards a consensus, which is the traditional idea behind the call for deliberative processes. In that small way, processes of societal *de facto* agenda-building and articulation of actor strategies are mimicked. The effect is not consensus, but increased reflexivity.

Notes

1 'Mature technological systems – cars, roads, municipal water supplies, sewers, tele-
phones, railroads, weather forecasting, buildings, even computers in the majority of
their uses – reside in a naturalized background, as ordinary and unremarkable to us as trees,
daylight, and dirt. Our civilizations fundamentally depend on them, yet we notice them
mainly when they fail, which they rarely do. They are the connective tissues and the
circulatory systems of modernity. In short, these systems have become infrastructures'
(Edwards 2003: 185).

2 Foucault (1977), Appadurai (1990) on 'technoscapes', Barry (2001) on 'technological zones of circulation'. Barry (2001: 200) comments, 'Foucault's analysis of *dispositifs* or apparatuses is too static to reveal the dynamic instability of socio-technical
arrangements'.

3 Other visualizations are possible, such as the fitness landscape (Lansing and Kremer
1993, Jelsma 2003), the epigenetic landscape with its 'chreodes' (Waddington 1975),
and a potential field, as in electromagnetic theory.

4 'Arenas and fora, and the various issues discussed and addressed there, thus involve . . .
political activity but not necessarily legislative bodies and counts of law' (Strauss
1978: 124).

5 The notion of entanglement is important, in order to avoid too-easy recourse to traditional interest and power explanations.

6 Implementation studies have gone some way in this direction when emphasizing the
importance of 'bottom-up' processes (Hanf and Toonen 1985); cf. also Pressman and
Wildavsky (1984) on mutual adaptation between policy making and what happens 'on
the ground', and who turned it into advice for modest policy making, or better, policy
making that takes implementability into account. In other words, goals are modified by
considering possible implementation.

7 Richard Jones made this remark in the Stanford-Paris conference on Social and Ethical
Implications of Nano- Bio-Info Convergence, Avignon, 18–19 December 2006. He
agreed to our quoting him this way.

8 Interestingly, the Swiss research institute Swiss Federal Laboratories for Materials Test-
ing and Research (EMPA) (see Merz in this volume), which has moved from materials
science and technology into nanotechnology, in a 2007 booklet *Reise in die Welt des
Nanometers* that explained nanotechnology to the general public, is prepared to say,
'Formuliert wurden diese Visionen [of molecular manufacturing and molecular self-
organization] erstmals 1981 von Eric Drexler. Er hat, 22 Jahre nach dem denkwürdigen
Vortrag Richard Feynmans, dessen nanowissenschaftliche Vision aufgenommen und zu
einer Vision Nanotechnologie weiterentwickelt. Heute gelten Eric Drexler, gemeinsam
mit Heinrich Roher und Gerd Binnig, die im selben Jahr das Rastertunnelmikroskop
erfanden, als die Väter der Nanotechnologie'. Nanoscientists at EMPA told us this text
was the responsibility of EMPA's communications department, not theirs.

9 Note the difficulty of terminology: terms like 'molecular machines' or 'assembly' and
'self- assembly' have been used (and thus claimed) by different parties, for different
purposes, and thus with different meanings. 'Molecular machines' is now a respectable
research area with concrete findings, and the researchers eschew any reference to the
Drexlerian use of molecular machines. 'Self-assembly' is sometimes used to refer to
Drexlerian replicators assembling copies of them- selves, but chemists from Whitesides (1995) on have claimed the term for what a 'society of molecules' can be induced
to do, rather than the precise control of atoms/molecules envisaged by Drexler (cf. also
Bensaude-Vincent 2006).

10 We are indebted to Colin Milburn for offering insights (and references) into the nature
of the early debate.

11 *The Globe and Mail* of 26 November 2002, reporting on the debut of the novel, also
quoted nanoscientist Wolfgang Heckl: 'We have to take this seriously. If enough

senators in the US get phone calls from their constituents saying, "I just read *Prey* and I'm scared," it could have a real impact on our funding. Nanoscience is just in its infancy. We can't afford to be cut off'. Interestingly, the Drexlerians were also concerned about loss of credibility, cf. how Chris Phoenix (Center for Responsible Nanotechnology) took the same (and misguided) approach of criticizing the science in *Prey* in his review in *Nanotechnology Now* (Phoenix 2003).

12 Nanotech promoters appear to have overestimated the extent to which the notion of Grey Goo would capture the public's imagination and evoke resistance against nanotechnology (Thurs and Hilgartner 2005). In fact, a 2004 Internet search by ETC Group indicated that most entries referring to the 'threat of Grey Goo as presented by Drexler and Crichton' were from nanotech promoters and scientists concerned over the alleged 'public misunderstanding of nanotechnology' that was assumed to be the result of earlier publicity on the notion of Grey Goo (ETC 2004: 7).

13 Here, we move away from Mario Kaiser's diagnosis that 'there is hardly any doubt that concerns such as the possible future existence of grey goo have initiated a somewhat vehement reflection on the foundations that nanoscience and technology rest upon' (Kaiser 2006: 5). As we see it, Smalley (and also George Whitesides) took a chemist's view of the matter and criticized Drexler's engineering vision on that basis. The Grey Goo scenario is referred to only in passing. Bennett and Sarewitz (2006: 315) also emphasize such a link: the need to avoid Bill Joy's conclusion that certain lines of investigation should be relinquished (e.g. self-replication of nanobots, which might spread to current work in nanotechnology).

14 There were other bones of contention, like human enhancement (artificial intelligence which exceeds human capacity), but these are not linked to Drexler's visions (Fisher and Mahajan 2006: 11).

15 A study 'to develop, insofar as possible, a consensus on whether molecular manufacturing is technically feasible'. And if feasible, the study would find 'the estimated time frame in which molecular manufacturing may be possible on a commercial scale; and recommendations for a research agenda necessary to achieve this result' (quoted from Regis 2004).

16 Peterson (2004: 12), vice president of the Foresight Institute, refers to successful lobbying of opponents to the molecular manufacturing vision. See also Tim Harper's comments (TNT Weekly 2003).

17 Milburn (ibid.: 122) then argues, 'This rhetoric thoroughly deconstructs itself in a futile struggle for boundary articulation that has already been lost'. For all practical purposes, however, from 2004 onward, the boundary was maintained without much effort through general acceptance of the claim that the Drexlerian vision was just speculation.

18 Drexler himself articulated this dynamic. Brown (2001) reported that Drexler said that many scientists eagerly slapped the term 'nanotechnology' on their research when it was viewed as 'sexy', but became 'a little upset to find that they had a label on their work that was associated with outrageous, science-fictiony sounding claims about the future and scary scenarios and other things. . . . What nanoscale technologist would want the burden of such fears?' (Drexler 2004).

19 Or sometimes EHS, cf. *The Economist*, A little risky business. November 22nd, 2007.

20 Interview (by Marloes van Amerom, 7 July 2006) with Vicky Colvin, Director CBEN.

21 There are further indications, for example the lack of reference to nanoparticle risks in the Delphi study into benefits and potential drawbacks of using nanotechnology for health, commissioned by the German Bundesministerium für Bildung und Forschung, September 2002.

22 When a product featured as 'nano' turns out to have health effects (as happened in April 2006 with the German bathroom cleaner Magic Nano), the first concern is about damage to the image of nano (and everybody was relieved that – this time – it was the aerosol in the can, not nanomaterials that were responsible for users' health problems; there may not even have been a nanomaterial in the product).

23 The IRGC is a private not-for-profit foundation, based in Geneva, 'to support govern-
    ments, industry, NGOs and other organizations in their efforts to understand and deal
    with major and global risks facing society and to foster public confidence in risk gov-
    ernance' (quoted from Renn and Roco 2006b: 5). A conference report is available from
    Swiss Re Centre for Global Dialogue (2007).
24 This framework is criticized by other NGOs; see the Open Letter by the Civil Society –
    Labour Coalition of 12 April 2007.
25 Degussa's website on nanotechnology has an item to this extent on responsibility
    (www.degussa- nano.com/nano (accessed on January 27, 2008)), and BASF's Code of
    Conduct has a similar thrust.
26 See: www.responsiblenanocode.org (accessed on January 27, 2008).
27 Even if the possibility of molecular manufacturing is kept alive in scenario-building
    exercises by the Center for Responsible Nanotechnology (CRN 2007), and in argu-
    ments about prudent anticipation (Lin and Althoff 2007).

References

Abbott, A. (1988), *The System of Professions*, Chicago: University of Chicago Press.

Albert de la Bruhèze, A. A. (1992), *Political Construction of Technology: Nuclear Waste Disposal in The United States 1945–1972*, PhD Thesis, Enschede: University of Twente.

Allis, D. G. (2007), *Interview With Sander Olson*, www.somewhereville.com/?p=93 (accessed on January 8, 2008).

Anderson, A., S. Allan, A. Petersen and C. Wilkinson (2005), 'The Framing of Nanotech-nologies in the British Newspaper Press', *Science Communication* 27(2): 200–220.

Appadurai, A. (1990), 'Disjuncture and Difference in the Global Cultural Economy', *Theory, Culture and Society* 7(2–3): 295–310.

Arnall, A. H. (2003), *Future Technologies: Today's Choices: Nanotechnology, Artificial Intelligence and Robotics: A Technical, Political and Institutional Map of Emerging Technologies*, London: Greenpeace Environmental Trust.

Barry, A. (2001), *Political Machines: Governing a Technological Society*, London and New York: The Athlone Press.

Baum, R. (2003), 'Nanotechnology: Drexler and Smalley Make the Case For and Against Molecular Assemblers', *Chemical Engineering News* 81(48): 37–42.

Bennett, I. and D. Sarewitz (2006), 'Too Little, Too Late? Research Policies on the Societal Implications of Nanotechnology of Nanotechnology in the United States', *Science as Culture* 15(4): 309–325.

Bensaude-Vincent, B. (2006), 'Two Cultures of Nanotechnology?', in J. Schummer and D. Baird (eds.), *Nanotechnology Challenges: Implications for Philosophy, Ethics and Society*, Singapore: World Scientific Publishing: 7–28.

Berube, D. and J. D. Shipman (2004), 'Denialism: Drexler vs. Roco', *IEEE Technology and Society Magazine* 23(4): 22–26.

Brown, D. (2001), 'Drexler Warns Terror Symposium: Nanotechnology Has Extreme Downsides', *Small Times* (19 December), www.smalltimes.com/document_display. cfm? document_id=2754.

Browne, W. R. and B. L. Feringa (2006), 'Making Molecular Machines Work', *Nature Nanotechnology* 1: 25–35.

Brumfiel, G. (2003), 'A Little Knowledge', *Nature* 424: 246–248.

Colvin, V. L. (2003), *Testimony of Dr. Vicki L. Colvin Director Center for Biological and Environmental Nanotechnology (CBEN) and Associate Professor of Chemistry Rice University, Houston, Texas Before the U.S. House of Representatives Committee on*

Science in regard to 'Nanotechnology Research and Development Act of 2003', 9 April, http://gop.science.house.gov/hearings/full03/apr09/colvin.htm (accessed on January 27, 2008).

CRN (2007), *The Center for Responsible Nanotechnology Scenario Project*, http://crnano. org/CTF-ScenarioIntro.htm (accessed on January 27, 2008).

Drexler, K. E. (1986), *Engines of Creation: The Coming Era of Nanotechnology*, New York: Anchor Press/Doubleday.

Drexler, K. E. (2004), 'Nanotechnology: From Feynman to Funding', *Bulletin of Science, Technology & Society* 24(1): 21–27.

Edwards, P. (2003), 'Infrastructure and Modernity: Force, Time, and Social Organization in the History of Sociotechnical Systems', in T. J. Misa, P. Brey and A. Feenberg (eds.), *Modernity and Technology*, Cambridge, MA: MIT Press: 185–225.

ETC (2003a), *The Big Down: Atomtech: Technologies Converging at the Nano-Scale*. www.etcgroup.org/en/materials/publications.html?pub_id=171 (accessed on October 10, 2008).

ETC (2003b), 'Size Matters! The Case for a Global Moratorium', *Occasional Paper Series* 7(1), www.etcgroup.org/en/materials/publications.html?id=165 (accessed on October 10, 2008).

ETC (2003c), 'Nanotechnology Un-gooed! Is the Grey/Green Goo Brouhaha the Industry's Second Blunder?', *Group Communiqué* 80, www.etcgroup.org/upload/publication/154/01/ecom_prince007final.pdf (accessed on October 10, 2008).

ETC (2004), 'Nanotechnology News in Living Color: An Update on White Papers, Red Flags, Green Goo, Grey Goo (and Red Herrings)', *ETC Group Communiqué* 85, www.etcgroup. org/en/materials/publications.html?pub_id=95 (accessed on October 10, 2008).

Feder, B. (2003), 'Nanotechnology Creates a Royal Stir in Britain', *International Herald Tribune* (May 20).

Fisher, E. and R. L. Mahajan (2006), 'Contradictory Intent? US Federal Legislation on Integrating Societal Concerns into Nanotechnology Research and Development', *Science and Public Policy*, 33(1): 5–16.

Foucault, M. (1977), *Discipline and Punish*, Harmondsworth: Penguin.

Guzman, K. A. D., M. R. Taylor and J. F. Banfield (2006), 'Environmental Risks of Nano-technology: National Nanotechnology Initiative Funding, 2000–2004', *Environmental Science & Technology* 40: 1401–1407.

Hanf, K. and T. A. J. Toonen (eds.) (1985), *Policy Implementation in Federal and Unitary Systems*, Dordrecht: Martinus Nijhoff Publishers.

Harper, T. (2006), 'Keynote', BizTech 2006 Meeting, Eindhoven (September 26).

Hilgartner, S. H. and B. V. Lewenstein (2005), 'Nugget', in *Project Description Nanotechnology as a "Revolutionary Technology": Rhetoric, Forward-Looking Statements, and Public Understanding of Science*, www.cns.cornell.edu/TechnologyAndSociety.html.

HRH The Prince of Wales (2004), 'Menace in the Minutiae: New Nanotechnology Has Potential Dangers as Well as Benefits', *The Independent on Sunday* (11 July): 3, 25.

Institute of Physics (2004), *Nanotechnology: Fear or Fiction? Explaining the Science: Identifying the Issues*, www.iop.org/activity/policy/Events/Lectures/file_5756.doc (accessed on August 1, 2008).

Jelsma, J. (2003), 'Innovating for Sustainability: Involving Users, Politics and Technol-ogy', *European Journal of Social Science Research* 16(2): 103–116.

Kaiser, M. (2006), 'Drawing the Boundaries of Nanoscience: Rationalizing the Concerns?', *Journal of Law, Medicine & Ethics* 34(4): 667–674.

Kearnes, M. and A. Rip (2009), 'The Emerging Governance Landscape of Nanotechnology', in S. Gammel, A. Lösch and A. Nordmann (eds.), *Jenseits von Regulierung: Zum politischen Umgang mit Nanotechnologie*, Berlin: Akademische Verlagsanstalt.

Kingdon, J. W. (1984), *Agendas, Alternatives and Public Policies*, Boston/Toronto: Little, Brown and Company.

Lansing, J. S. and J. N. Kremer (1993), 'Emergent Properties of Balinese Water Temple Networks: Coadaptation on a Rugged Fitness Landscape', *American Anthropologist* 95(1): 97–114.

Lin, P. and S. Althoff (2007), 'Nanoscience and Nanoethics: Defining the Disciplines', in F. Allhof, P. Lin, J. Moor and J. Weckert (eds.), *Nanoethics: The Ethical and Social Implications of Nanotechnology*, Hoboken, NJ: John Wiley & Sons: 3–16.

Los Angeles Times (2002), 'The Future Dances on a Pin's Head; Nanotechnology: Will It be a Boon: Or Kill Us All?', *Los Angeles Times* (November 26).

Lovy, H. (2003a), *NanoBusiness Leader Makes the Call* (December 9), http://nanobot. blogspot.com/2003/12/nanobusiness-leader-makes-call.html (accessed on March 4, 2009).

Lovy, H. (2003b), *Drexler on 'Drexlerians'* (December 15), http://nanobot.blogspot. com/2003/12/drexler-on-drexlerians.html (accessed on March 4, 2009).

Lux Research (2005), 'A Prudent Approach to Nanotechnology Environmental, Health, and Safety Risks', *Industrial Biotechnology* 1(3): 146–149.

Maynard, A. D., R. J. Aitken, T. Butz, V. Colvin, K. Donaldson, G. Oberdörster, M. A. Philbert, J. Ryan, A. Seaton, V. Stone, S. S. Tinkle, L. Tran, N. J. Walker and D. B. Warheit (2006), 'Safe Handling of Nanotechnology', *Nature* 44: 267–269.

Menon, J. (2004), *Swiss Re Warns Insurers on Nanotechnology After Study (Update 2)* (May 10), www.bloomberg.com/apps/news?pid=10000085&sid=aDKqgCoPlBPI&refer=europe (accessed on March 4, 2009).

Milburn, C. (2004), 'Nanotechnology in the Age of Posthuman Engineering: Science Fiction as Science', in N. K. Hayles (ed.), *Nanoculture: Implications of the New Technoscience*, Bristol, UK: Intellect Books: 109–129 and 217–223. Reprinted from *Configurations* 10(2): 261–296.

Milburn, C. (2008), *Nanovision: Engineering the Future*, Durham: Duke University Press.

Munich Re Group (2002), *Nanotechnology: What Is in Store for Us?* www.munichre.com/ publications/302–03534_en.pdf?rdm=40215 (accessed on January 15, 2008).

National Science and Technology Council (NSTC) and Interagency Working Group on Nanoscience, Engineering and Technology (IWGN) (1999), *Nanotechnology: Shaping the World Atom by Atom*, Washington, DC: NSTC.

Nordan, M. M. and M. W. Holman (2005), 'A Prudent Approach to Nanotechnology Environmental, Health and Safety Risks', *Industrial Biotechnology* 1(3): 146–149.

Nowotny, H., P. Scott and M. Gibbons (2001), *Re-Thinking Science: Knowledge and the Public in an Age of Uncertainty*, Cambridge, UK: Polity Press.

Oberdörster, G., E. Oberdörster and J. Oberdörster (2005), 'Nanotoxicology: An Emerging Discipline Evolving From Studies of Ultrafine Particles', *Environmental Health Perspectives* 113(7): 823–839.

Oliver, J. (2003a), 'Charles: "Grey Goo" Threat to the World', *The Mail on Sunday* (April 27): 1.

Oliver, J. (2003b), 'Nightmare of the Grey Goo', *The Mail on Sunday* (April 27): 8.

Peterson, C. L. (2004), 'Nanotechnology: From Feynman to the Grand Challenge of Molecular Manufacturing', *IEEE Technology and Society Magazine* 23(4): 9–15.

Phoenix, C. (2003), 'Don't Let Crichton's Prey Scare You: The Science Isn't Real', *Nanotechnology Now* (January), www.nanotech-now.com/Chris-Phoenix/prey-critique.htm (accessed on January 15, 2008).

Phoenix, C. and E. Drexler (2004), 'Safe Exponential Manufacturing', *Nanotechnology* 15: 869–872, www.iop.org/EJ/abstract/0957-4484/15/8/001/ (accessed on January 15, 2008).

PhysicsOrg.com (2005), *More Funds* (November), www.physorg.com/news4963.html

Pressman, J. L. and A. Wildavsky (1984), *Implementation: How Great Expectations in Washington Are Dashed in Oakland*, third edition, Berkeley: University of California Press.

Regis, E. (2004), 'The Incredible Shrinking Man', *Wired* (October), www.nanotechnolo gist.com/misc/index.html (accessed on January 15, 2008).

Renn, O. and M. C. Roco (2006a), 'Nanotechnology and the Need for Risk Governance', *Journal of Nanoparticle Research* 8(2): 153–191.

Renn, O. and M. C. Roco (2006b), *Nantechnology Risk Governance*, White Paper No. 2, Geneva: IRGC.

Rincon, P. (2004), 'Nanotechnology Guru Turns Back on "Goo"', *BBC News* (June 9), http://news.bbc.co.uk/1/hi/sci/tech/3788673.stm (accessed on January 15, 2008).

Rip, A. (1986), 'Controversies as Informal Technology Assessment', *Knowledge* 8(2): 349–371.

Rip, A. (2006a), 'The Tension Between Fiction and Precaution in Nanotechnology', in E. Fisher, J. Jones and R. von Schomberg (eds.), *Implementing the Precautionary Principle: Perspectives and Prospects*, Cheltenham/Northhampton, MA: Edward Elgar.

Rip, A. (2006b), 'A Coevolutionary Approach to Reflexive Governance: And Its Ironies', in J.-P. Voß, D. Bauknecht and R. Kemp (eds.), *Reflexive Governance for Sustainable Development*, Cheltenham/Northhampton, MA: Edward Elgar: 82–100.

Rip, A. (2006c), 'Folk Theories of Nanotechnologists', *Science as Culture* 15(4): 349–365.

Rip, A., D. Robinson and H. te Kulve (2007), 'Multi-Level Emergence and Stabilisation of Paths of Nanotechnology in Different Industries/Sectors', *Workshop on Paths of Developing Complex Technologies*, Berlin, September 16–17.

Rip, H. and H. te Kulve (2008), 'Constructive Technology Assessment and Sociotechnical Scenarios', in E. Fisher, C. Selin and J. M. Wetmore (eds.), *The Yearbook of Nanotechnology in Society, Volume I: Presenting Futures*, Berlin etc: Springer: 49–70.

Robinson, D., M. Ruivenkamp and A. Rip (2007), 'Tracking the Evolution of New and Emerging S&T via Statement-Linkages: Vision Assessment of Molecular Machines', *Scientometrics* 70(3): 831–858.

Roco, M. C. (2003), 'Broader Societal Issues of Nanotechnology', *Journal of Nanoparticle Research* 5(3–4): 181–189.

Roco, M. C. and W. Bainbridge (2001), *Societal Implications of Nanoscience and Nanotechnologies*, Boston: Kluwer Academic Publishers.

Roco, M. C. and R. Tomellini (eds.) (2003), 'Nanotechnology: Revolutionary Opportunities and Societal Implications', *EC – NSF Workshop* (January 2003), Luxembourg: European Communities.

Royal Society/Royal Academy of Engineering (2004), *Nanoscience and Nanotechnologies: Opportunities and Uncertainties*, London: Royal Society. www.nanotec.org.uk/report/contents.pdf (accessed on January 27, 2008).

Sahal, D. (1985), 'Technological Guideposts and Innovation Avenues', *Research Policy* 14: 61–82.

Sample, I. (2004), 'Civilisation Safe as Nanobot Threat Fades', *Guardian* (June 9), www.guardian.co.uk/uk_news/story/0,3604,1234436,00.html (accessed on January 27, 2008).

The Scotsman (2004), *Grey Goo Destruction Theorist Changes Tack* (June 8), http://news. scotsman.com/weirdoddandquirkystories/Grey-goo-destruction-theorist-changes.2536010. jp (accessed on March 4, 2009).

Selin, C. (2007), 'Expectations and the Emergence of Nanotechnology', *Science, Technology & Human Values* 32(2): 196–220.

Sherriff, L. (2004), 'World Safe From Nanobot "Grey Goo": U-turn by Prophet of Doom Eric Drexler', *The Register* (June 9), www.theregister.co.uk/2004/06/09/grey_goo_ goeth/ (accessed on January 27, 2008).

Shibuya, E. (1996), 'Roaring Mice Against the Tide: The South Pacific Islands and Agenda-Building on Global Warming', *Pacific Affairs* 69.

Small Times (2002), 'Activists: No More Nanomaterials Until We Know Whether It Pollutes' (8 August), www.smalltimes.com/articles/stm_print_screen.cfm?ARTICLE_ ID=268070 (accessed on January 27, 2008).

Small Times (2003a), 'Watchdogs Say Stop Nanotech, Start Worldwide Dialogue' (31 January), www.smalltimes.com/articles/stm_print_screen.cfm?ARTICLE_ID=268506 (accessed on January 27, 2008).

Small Times (2003b), 'Greenpeace Wades into Nanodebate With Report that Calls for Caution' (24 July), www.smalltimes.com/Articles/Article_Display.cfm?ARTICLE_ ID=268886&p=109 (accessed on March 4, 2009).

Smalley, R. (2001), 'Of Chemistry, Love and Nanobots', *Scientific American* 285: 76–77.

Smalley, R. (2003), 'Smalley Concludes', in Baum, R. (2003), Nanotechnology, Drexler and Smalley Make the Case For and Against Molecular Assemblers, *Chemical Engineering News*, 81(48), pp. 37–42, at p. 42.

Stirling, A. (2005), 'Opening Up or Closing Down: Analysis, Participation and Power in Social Appraisal of Technology', in M. Leach, I. Scoones and B. Wynne (eds.), *Science, Citizenship and Globalisation*, London: Zed Books: 218–231.

Strauss, A. (1978). 'A Social World Perspective', *Studies in Symbolic Interaction* 1: 119–128.

Swierstra T. and A. Rip (2007), 'Nano-ethics as NEST-ethics: Patterns of Moral Argumentation about New and Emerging Science and Technology', *NanoEthics* 1(1): 3–20.

Swiss Re (2004), *Nanotechnology: Small Matter, Many Unknowns*, Zürich: Swiss Reinsurance Company (Swiss Re).

Swiss Re Centre for Global Dialogue (2007), 'The Risk Governance of Nanotechnology: Recommendations for Managing a Global Issue', *Conference Report*, July 6–7, 2006, Zürich: Swiss Reinsurance Company.

Thurs, D. and S. Hilgartner (2005), *The Spread of Grey Goo: Fearful Publics and Fear of the Public in the Nanotechnology Arena*, www.cns.cornell.edu/documents/TheSpre adofGreyGooMarch05.pdf

TNT Weekly (2003), 'The Plot Thickens and the Nanotechnology Bill Gets Sillier', www. cientifica.info/html/TNT/tnt_weekly/archive_2003/Issue_13.htm (accessed on January 28, 2008).

Trubek, D. M., P. Cottrell and M. Nance (2006), 'Soft Law, Hard Law and EU Integration', in G. De Burca and J. Scott (eds.), *Law and New Governance in the UK and the US*, Oxford: Hart Publising: 65–94.

Van Noorden R. (2007), 'Building Tomorrow's Nanofactory', *Chemistry World* (October 19), www.rsc.org/chemistryworld/News/2007/October/19100701.asp (accessed on January 27, 2008).

Waddington, C. H. (1975), *The Evolution of an Evolutionist*, Ithaca, NY: Cornell University Press.

Whitesides, G. M. (1995), 'Self-Assembling Materials', *Scientific American* (September): 146–149.

Whitesides, G. M. (1998), 'Nanotechnology: Art of the Possible', *Technology: MIT Magazine of Innovation* (November/December): 8–13.

Whitesides, G. M. (2001), 'The Once and Future Nanomachine', *Scientific American* 285(3): 78–83.

Wood, S., A. Geldart and R. Jones (2007), *Nanotechnology: From the Science to the Social*, Swindon, UK: Economic and Social Research Council.

6 Positions and responsibilities in the 'real' world of nanotechnology

Arie Rip and Clare Shelley-Egan

Published as

Positions and responsibilities in the 'real' world of nanotechnology, in René von Schomberg and Sarah Davies (eds.), *Understanding Public Debate on Nanotechnologies. Options for Framing Public Policy*, Brussels: Commission of the European Communities, January 2010, pp. 31–38

Introduction

The eventual performance and application of emerging technologies such as nanotechnology is uncertain, and their further effects on society are even more uncertain. Still, visions are put forward and debated and actions are taken. The German sociologist Ulrich Beck has diagnosed contemporary society as showing 'organised irresponsibility': modern technological society allows scientists, engineers, and industry to develop and introduce all sorts of new technologies (chemical, nuclear, genetic modification) while it structurally lacks means to hold anyone accountable (Beck 1988, 1995). As Merkx (2008) has argued, it would be better to think in terms of an ongoing organization of responsibilities, which cannot always keep up with advances in science, technology, and industry. For nanotechnology, there is recognition of this problem and there are attempts to articulate what responsible development might be (there is, in other words, reflexivity). However, it is not always clear what this implies for organizing responsibilities.

Accordingly, we have studied what present visions and attempts at articulating responsibilities are by looking at ethics in, as it were, the 'real' world. A striking fact is that nanoscience and nanotechnology are still so open and uncertain that there are almost no specific ethical issues and challenges. The various actors involved fall back on their own positions and what they see as their 'mandate' to justify their visions and actions. This is understandable as a way to reduce complexity.

Briefly, in our interviews we saw how scientists drew on a standard repertoire in which science and ethics are separate: they had 'recourse to the technical' and alluded to their (partly self-defined) mandate to work towards progress in science. Industrial actors, and chemical companies in particular, were concerned about lack of trust in industry and showed enlightened self-interest in their involvement in initiatives around the responsible development of nanotechnology. For NGOs, there may be a standard repertoire as well, about the need for concern and to be precautionary. (We note in passing that publics as such have no positions/mandates, so they fall back on general cultural repertoires when asked to say something. See Davies, Macnaghten and Kearnes 2009; Ferrari and Nordmann 2009.)

These standard repertoires build on what we call (present) divisions of moral labour and allow the actors to continue to play their roles. For example, the repertoires of scientists and industrialists reflect an 'enactor' perspective: the promise of nanotechnology must be pushed and ethics is seen as a brake on progress. While some division of moral labour is necessary, one should not assume that the present division, with its roots in the past, is still adequate. To enable critical reflection, by the enactors themselves as well as by other commentators, it is important that standard repertoires are opened up.

In this chapter we will develop this diagnosis on the basis of the presentation and analysis of interview and other data.

Positions and visions

The 'real' world is full of references to accepted and/or desired roles and responsibilities of various actors. This in and of itself includes ethical stances. The reference to 'responsible development' induces further positioning. An indicative (and strongly formulated) example is offered by US Under-Secretary of Commerce Philip Bond:

> Given nanotechnology's extraordinary economic and societal potential, it would be unethical, in my view, to attempt to halt scientific and technological progress in nanotechnology. . . . Given this fantastic potential, how can our attempt to harness nanotechnology's power at the earliest opportunity – to alleviate so many earthly ills – be anything other than ethical? Conversely, how can a choice to halt be anything other than unethical?
>
> (Bond 2004)

While scientists and industrialists might subscribe to such a vision in general, their immediate concerns are more mundane.

In a small focus group, a scientist said,

> Well, basically with my work, I would say my duty includes having a prosperous group . . . which is good for the university, so the university wants me to do this, then yeah, this duty requires me to . . . look at what funding agencies are asking for.

The issue of responsibility was not always reduced in such a clear-cut way. The tension between developing a technology and the use to which the technology is put was seen by the participants as a grey area in the responsibility of scientists:

If you invent something, you are responsible for the invention and it has certain consequences – there are various applications and you cannot predict all these applications. It can be misused in the future and then the question is – are you still responsible for that as a researcher?

A scientist we interviewed on the same topic was very clear about who should be responsible for applications:

I think it's mainly industry and people who are selling products and putting them onto the market who should be asked OK, is this safe or not, and they can ask us to help them to answer this question.

Clearly, these are attempts to invoke a division of moral labour where it is clear who should do what and who should be held responsible. In their study of a year-long interaction between bioscientists and citizens in Vienna, Felt and Fochler noted how

Scientists . . . more or less explicitly rejected taking responsibility for any consequences of the knowledge they produced beyond quite narrowly defined imminent risks arising from their work (such as transgenic mice escaping from their lab). Taking any role in the governance of these consequences was not part of how they envisaged their professional role. Their argument for not doing so was to be only doing basic research, without any concrete focus on application. Any applications would need to be developed by other actors at a later stage, and such implications would then have to be decided 'by society'.

(Felt and Fochler 2008: 495)

In the exercise on which Felt and Fochler reported, the citizens did not accept this division of moral labour:

While this issue was not that disturbing for the scientists, because they only felt marginally concerned by this question, many citizens were quite upset by the scientists' refusal to consider their responsibility as an issue to reflect on.

(Felt and Fochler 2008: 495)

Our point is not to criticize this stance of scientists – there might be good reasons for such a division of labour. We want to ask, though, whether this vision is sufficiently thought through and whether it goes beyond referencing a standard repertoire. This question is pertinent because there is also a tendency of scientists to cash in on the good things, but refuse responsibility for the bad things. As Jerry Ravetz once formulated this dynamic, 'Scientists take credit for penicillin, but Society takes the blame for the Bomb' (Ravetz 1975: 46).

Industrialists can play a similar game of praise and blame when they put all responsibility on consumers who buy (or do not buy) products, or, more abstractly, refer to market forces which are outside of their control. The industrialists in our interviews were more focussed, however. When asked about responsible innovation, interviewees said (with reference to corporate social responsibility),

> That's part of our DNA.
>   It's part of the total philosophy . . . it's a total attitude – you can't just split parts of it . . . it's part of the total way we do business.

In particular, chemical companies, with their experience of the so-called Responsible Care Program (which commits to working in an environmentally friendly way), were willing to take up notions of responsible development. In practice, this means attention to safety and health issues of employees and transparency to the outside world.

In general, it can be an asset for a firm to be seen as a 'good firm'. A respondent from a chemical company said,

> Our approach to transparency is as a result of our experiences in other technology and safety debates . . . it's a kind of lesson learned and what we now want to do is to go another way, to go into the public debate of nanotech. And we hope it will end more successfully than other debates. So, we do this very early, this is sometimes also difficult but for the moment we see no alternative to go another way.

This kind of early engagement with nanotechnology is double-edged, however:

> It is a risk and sometimes ends in reputational damage [of your company]. Those companies that are transparent are also the focus of NGO debates because nothing is known [of what other companies are doing].

There is an explicit ethics of enlightened self-interest here.

For broader issues, such as the debate on precaution and a possible (partial) moratorium – for example on nanoparticles, as called for by NGOs like ETC group and Friends of the Earth – broader views were offered. There was a unified negative response to the call for a moratorium. Respondents felt that a moratorium would stop progress, and referred to the benefits which nanotechnology can bring to mankind and the environment. Their earlier reference to gradual evolution ('we're continuing what we were doing already') shifted and they now positioned nanotechnology as 'revolutionary', referring to its role in the effort towards climate protection or in the fight against cancer.

Interestingly, the industrialists' view of the role of NGOs referred to a division of moral labour, starting with a practical observation. One respondent saw the call for a moratorium as 'a bit of a knee-jerk reaction' but conceded that 'they're right in one sense, I guess, there's always a chance that we don't understand [the risks]'.

This can be elaborated, as when another respondent said he thought the concern of NGOs about nanotechnology is

A very good thing, in the sense that there are groups of people who watch the developments and look critically at it, ask questions to make sure that everybody is keen on, let's say, the balance between opportunities and the potential risks. Well, that's the impression I have. Also, I believe that even the groups that are sort of aware and ask critical questions, my personal impression is that they are also looking for the balance and about what is really the issue; and only the ones who are very political will make a firm statement like there should be a ban before we know enough.

Here, the respondent introduces a distinction between 'good' and 'bad' NGOs. The latter category appears to be upfront in the mind of industrialists. In meetings and interviews, one can actually hear industrialists (and other enactors) qualifying such bad NGOs as agitators, failing to act in good faith, using misleading information to further their cause and painting different nanotechnologies with the same brush.

Divisions of moral labour

Division of moral labour is a general phenomenon in our society. It refers to a division of obligations and commitments, as well as to notions regarding who is eligible to be praised or blamed. The division of moral labour can be approached normatively by inquiring into whether the present division of moral labour in science is a 'good' one. This is also an ethical (and political) question, but of a second-order. It makes visible ethical, and more broadly normative, aspects by inquiring into the justification of present arrangements, rather than taking them for granted. One example is the justification of the common division of moral labour for science: that scientists have a moral obligation to work towards progress and that that is how they discharge their duty to society. This mandate is legitimate as long as scientists deliver on what they promised, while maintaining the integrity of science.

While aspects of the division of moral labour in the nanotechnology world can be criticized, the solution is not to dispense with the division of moral labour. The challenge of organizing responsibilities remains.

At the moment, one sees recourse to traditional roles (by critical actors, such as some NGOs, as well as by scientists) and the use of standard repertoires. Moreover, there is a focus on concrete issues of transparency and risk, while the repertoires and divisions of labour that shape the debate on such issues are backgrounded. This may well create limitations in working towards productive solutions. Thus there is a need to critically reflect on standard repertoires and on the value of implied and *de facto* divisions of moral labour.

There is an important ethico-political point involved here. A division of moral labour is effective when it is accepted and implemented, as if it were solid – or, better, solidified. In changing circumstances (which might include changing values, for example on precaution or about participation), where responsibilities may have to be redefined, the solidity of the division of labour will become a hindrance rather than a help. It has to be opened up, 'melted' as it were, so that there is space for new configurations. (In this respect, the conceptual distinction introduced by Swierstra and Rip (2007) is useful: while morality is characterized by unproblematic acceptance, ethics is characterized by explicit discussion and controversy. Thus, ethics is carried out when moral routines and standard repertoires of moral argumentation are put up for discussion. Swierstra and Rip then refer to ethics as 'hot' morality and to morality as 'cold' ethics – highlighting the notions of melting and solidifying.)

Actually, one sees the opening up of existing divisions of moral labour in a number of ways.

First, recent initiatives for codes of conduct for nanotechnology and receptivity to these initiatives, in terms of willingness to discuss such codes seriously, provide an opportunity for the opening up of standard repertoires. A code of conduct is a self-binding action. A key feature of such codes is that they assume, and thus create, a public space where a subscriber to the code can be called to account by other actors referring to the code. This is reflected in the (justified) reluctance of industry and other actors to subscribe to broad circumscriptions in codes, as they open up to too many (unexpected, risky) critical calls for being held to account. On the other hand, such public spaces allow for deliberation and 'probing each other's worlds' and can be used for learning by all parties.

Second, there are 'third parties' who do not develop nanotechnology themselves but who exert leverage on developments through their actions. For our question of opening up, the entrance point is that third parties such as funding agencies (for science) and venture capitalists and insurance companies (for industry) may require anticipation on adequate societal embedding. There are already examples of such third parties becoming proactive. Nano-enactors must then develop relevant competencies and act upon them.

Codes of Conduct and proactive third parties can be seen as soft governance structures, offering direction but without authoritative sanctions. What is important here is not whether they are effective or not (because the goals they pursue may not be appropriate), but whether they stimulate critical reflection on background issues, and thus provide openings for longer-term changes.

Ethical reflexivity rather than consensus

It is clear that there are differences in the visions and ethical stances of actors, and these can be a challenge for communication and further interaction. This does not imply, however, that one should strive for consensus – definitely not at this stage of development, in which opening up earlier repertoires and roles is important. In addition, difference also provides opportunities. This implies that there is a need for 'design for difference' methodologies which can be implemented in order to handle difference and a variety of perspectives in interactions.

In such interactions and attempts to solidify elements of a new and better division of moral labour, much more is at stake than ethical reflexivity at the individual level.

In the real world, ethical reflexivity of actors is caught between individual agency and institutional role. Individual ethical reflection runs up against institutional and moral divisions of labour, both in the nature of the reflection and in the boundaries set on action based on such reflection. In other words, for change to occur, there have to be openings at the institutional level.

This is where Beck's (1994) diagnosis of 'reflexive modernisation' comes in. Institutions of modernity, including science, are confronted with the effects of what they do (how they function) and start to change, somewhat reluctantly. This introduces a reflexive component into their functioning. Broadening the notion of the ethical, one can speak of ethical reflexivity at the institutional level.

Actually, a further broadening is possible, by considering the politics of responsible innovation. There are of course immediate politics in the sense of struggles and negotiations about directions to go and arrangements to make. However, there are also 'deep' politics, in the sense that divisions of moral labour and, in the end, the constitution of late industrial society may shift. And this may happen reflexively.

Nanotechnology may not be as revolutionary in its impacts as some of its proponents present it. However, it may well be revolutionary in that it is the occasion to explicitly, and at an early stage, combine immediate and 'deep' politics. While the outcomes are unclear at this stage, the process is important and should be nurtured. Our analysis offers building blocks.

Bibliography

Beck, Ulrich (1988) *Gegengifte. Die organisierte Unverantwortlichkeit.* Frankfurt/Main: Suhrkamp Verlag.

Beck, Ulrich (1995) *Ecological Politics in an Age of Risk.* Cambridge: Polity Press.

Beck, Ulrich, Giddens, Anthony, and Lash, Scott (1994) *Reflexive Modernization.* Cambridge: Polity Press.

Bond, Philip J. (2004) *Responsible Nanotechnology Development,* in SwissRe Workshop, December. Zürich: Swiss Re.

Davies, Sarah, Macnaghten, Phil, and Kearnes, Matthew (eds.) (2009) *Reconfiguring Responsibility: Lessons for Public Policy (Part 1 of the Report on Deepening Debate on Nanotechnology).* Durham: Durham University.

Felt, Ulrike, and Fochler, Maximilian (2008) The bottom-up meanings of public participation in science and technology, *Science and Public Policy 35*(7), 489–499.

Ferrari, Arianna, and Nordmann, Alfred (2009) *Reconfiguring Responsibility: Lessons for Nanoethics (Part 2 of the Report on Deepening Debate on Nanotechnology).* Durham: Durham University.

Merkx, Femke (2008) *Organizing Responsibilities for Novelties in Medical Genetics.* Enschede: University of Twente. PhD thesis, defended 12 December 2008.

Ravetz, Jerome R. (1975) . . . et augebitur scientia, in Harré, Rom (ed.), *Problems of Scientific Revolution: Progress and Obstacles to Progress in the Sciences.* London: Oxford University Press, pp. 42–75.

Swierstra, Tsjalling, and Rip, Arie (2007) Nano-ethics as NEST-ethics: Patterns of moral argumentation about new and emerging science and technology, *NanoEthics 1*, 3–20.

7 De facto governance of nanotechnologies

Arie Rip

Published as

De facto governance of nanotechnologies, in Morag Goodwin, Bert-Jaap Koops and Ronald Leenes (eds.), *Dimensions of Technology Regulation*, Nijmegen: Wolf Legal Publishers, 2010, pp. 285–308.

Introduction

New and emerging technologies, especially nanotechnologies, with the structural uncertainties about their eventual functionalities and risks, are a challenge to governance. Regulatory agencies in Europe and the US review existing regulation and consider voluntary reporting as a transitional measure. Risk governance is opened up to include public dialogues and deliberative processes. What is striking is how much actual governance is already occurring in and around nanotechnology, without any particular actor being responsible for the emerging governance arrangements.

Thus, the first aim of this paper is to map what is happening: the actions and interactions and how these add up to outcomes at the collective level that function as governance arrangements. In that way, the paper is explorative: it is an attempt to understand what is occurring and is partly based on the author 'moving about' as a self-styled anthropologist in the world of nanotechnologies. What is clear is that the emerging governance arrangements have a distributed character. This is captured by using the notion of governance, which, in contrast to government, is distributed almost by definition. The additional point, however, is that bottom-up actions, strategies, and interactions are constitutive for these arrangements rather than that they are the result of an opening up of an earlier centralized arrangement to make it more distributed – a common way to introduce the notion of governance (Van Kersbergen and Van Waarden, 2004). To emphasize the strong bottom-up character of what is happening, I introduce the notion of *de facto* governance.

This leads to the second aim of this paper: to flesh out the notion of *de facto* governance by showing how it works in the domain of nanotechnologies. The recognition of the importance of *de facto* governance implies that attempts at regulation can be located as interventions in emerging *de facto* governance, and will depend on it for their effectiveness.[1] This is similar to the way in which Henry Mintzberg (1994) viewed intentional (and often top-down) strategy in firms and other organizations, noting that the latter's effects will depend on the interaction with *de facto*, or in his words 'pattern' and 'emergent' strategies, that are out there already. While society should not be seen as an organization writ large, the dual dynamics outlined by Mintzberg occur all the time. And they can add-up to what one could call a societal agenda.

This chapter has a third aim, linked to what I see as an intriguing potential *de facto* governance pattern: the internalization of requirements of 'responsible development' of nanosciences and nanotechnologies in the actual technological and product-development choices and strategies. Something of the sort is happening, as I will show, and the question then is what this implies for the governance of nanotechnologies. The further question is whether internalization of such considerations might occur for other emerging technologies as well. If this occurs, or is expected to occur, it will create a new regime for shaping technology development within society.

After fleshing out the notion of *de facto* governance, I will present and discuss two recent developments. Firstly, how a socio-technical agenda about promises and concerns over nanotechnology emerged, in which risks, and in particular risks of nanoparticles, became dominant. And secondly, how 'responsible development' has become an integral part of the discourse and, to some extent, of the practice, of nanotechnology. This then allows me to enquire into the possible internalization of societal considerations in ongoing development of nanosciences and nanotechnologies. In the concluding comments, I will come back to governance issues.

The notion of de facto governance

In the broadest sense of the concept of governance, all structuring of action and interaction that has some authority and/or legitimacy counts as governance. Authors such as Van Kersbergen and Van Waarden (2004) and Kooiman (2003) recognize this, even if they do not thematize it. Governance arrangements may be designed to serve a purpose, but can also emerge and become forceful when institutionalized. The same move is visible in Voß et al. (2006: 8) where they argue that governance refers to 'the characteristic processes by which society defines and handles its problems. In this general sense, governance is about the self-steering of society'. They then develop this further:

> Governance is understood as the result of interaction of many actors who have their own particular problems, define goals and follow strategies to achieve them. Governance therefore also involves conflicting interests and struggle for dominance. From these interactions, however, certain patterns emerge, including national policy styles, regulatory arrangements, forms of organisational management and the structures of sectoral networks. These patterns display the specific ways in which social entities are governed. They

comprise processes by which collective processes are defined and analysed, processes by which goals and assessments of solutions are formulated and processes in which action strategies are coordinated. . . . As such, governance takes place in coupled and overlapping arenas of interaction: in research and science, public discourse, companies, policy making and other venues.

This view has been offered before, notably by Elinor Ostrom. As Scharpf (1997: 204) phrases it,

> Much effective policy is produced not in the standard constitutional mode of hierarchical state power, legitimated by majoritarian accountability, but rather in associations and through collective negotiations with or among organizations that are formally part of the self-organization of civil society rather than of the policy-making system of the state.
>
> (Ostrom, 1990)

A specific aspect is highlighted by Braithwaite and Drahos (2000: 10), where they note,

> The global perspective on regulation we promote not only reframes individuals as subjects and objects of regulation (as in the drug case) and states as subject and object of regulation (by Moody's, the IMF, the Rothschilds and Greenpeace). Understanding modernity, we find, demands the study of plural webs of many kinds of actors which regulate while being regulated themselves.[2]

In such an encompassing view of governance, explicit attempts at steering intentional government arrangements will be seen as part and parcel of the overall process, not outside of it. In economics, one can speak of endogenizing a factor (like new technology) that had been considered as external to economic analysis. Similarly, one can now say that government and design of governance arrangements must be endogenized to capture what is happening (Rip, 2006; cf. also Voβ, 2007).

Where governance of technology is discussed, it appears to be reduced to either innovation stimulation or regulation of actual and possible side effects. The focus on performance of technologies (positive and/or negative) seems obvious, but it pushes to the background any broader consideration of governance. Science and Technology Studies (STS) have offered case studies and analysis showing that there is actually a lot of broader governance going on, but a lot of work is required to overcome the myopia of the prevalent view of technology being technically driven and/or naturalized. Moreover, as it is a prevalent view, the simple distinction between innovation stimulation and regulation is itself an example of a governance pattern in the broad sense.

This governance pattern derives from the gap between development and promotion of technology, and the responses of a society that emerged in the industrial revolution and stabilized in the nineteenth and early twentieth century (Rip et al., 1995). A version of the gap is visible in the institutional separation between promotion and control of new science and technology, for example the difference in outlook and activities between government departments for trade and industry on the one hand and for social, health, and environmental affairs on the other hand. To some extent this is a productive division of labour. But the separation of technology development (in firms, in public research institutes, in technical universities) from wider society implies that society has to respond, somehow, and is at a disadvantage because there have been investments in development already.[3] This 'gap' has led to an understanding of there being these two separate worlds, of 'enactors' of new technology vs. (civil) society, as well as an understanding that with regard to new technology, civil society is 'forced' into one or another of three reactions: to welcome it (this appears to occur for large parts of information and communication technology), to be fatalistic (for example about new infrastructural technologies), or to oppose it (as happened with agricultural biotechnology). The recent interest of technology 'enactors' in engaging civil society and having public dialogues on new technologies can be seen as an attempt to improve the possibility that society will respond in a welcoming rather than oppositional mode.

Another example of *de facto* governance arrangements, and one which has been highlighted by STS studies, are sociotechnical systems and infrastructures that together form the sociotechnical landscapes in which we live and move about. Roads and motorways serving automobile transportation, and the structures linked to them, are a clear example of how these 'arrangements' shape what we do and cannot do, and with the authority that comes with their being invisible because self-evidently 'given'.[4] Systems and infrastructures can have political effects, for example the material unification of a country like the Netherlands (Schot et al., 2003); see also Anderson (1991) for the role of sociotechnical regimes and the idea of a national community. The socio-technical landscapes in/of our societies are like a constitution, even if not drawn up by a constitutional assembly. This includes the disciplining (of actors) necessary to maintain them and have them develop in certain ways. Systems like mobile telephony, including infrastructure as well as evolving customs and rules of use, are further examples of emerging socio-technical regimes which function as a *de facto* constitution.[5]

For emerging sciences and technologies it is not yet clear what their possible sociotechnical constitutional effects might be, but one can anticipate them based on an understanding of dynamics of technological change and its embedding in society (Rip and Te Kulve, 2008). This requires what one might call non-linear thinking, especially for technologies like nanotechnology that are enabling technologies. That is, nanotechnology delivers new materials and components to help create better devices and systems, and it is the latter which deliver the desired functionalities, and thus shows sociotechnical agency. Thus, nanotechnology is said to just improve performance, and sometimes allow new functionalities (e.g. dirt-repellent surfaces) and should therefore not be an object of concern *qua* its societal effects. Still, nanotechnology could lead to major changes where certain thresholds are passed. For example, when RFID (Radio Frequency Identification Devices) becomes cheaper and smaller thanks to nanotechnology, and thus more widely available as well as better implantable, all products can be traced individually and an 'Internet of Things' becomes possible, as well as implants becoming easy and almost natural, leading to a view of the implantable and thus 'readable' human. All this is still to come, but it is being discussed already and may lead to governance measures and arrangements. One could call this anticipatory governance (Barben et al., 2007). In fact, there is an anticipatory component to all governance (Rip, 2006).

The role of technology in governance is one of solidifying arrangements by embodying them in material form, or as Pels et al. (2002: 2) phrase it, 'the performative and integrative capacity of "things" to help make what we call society'. In the case of the overpasses on Long Island (see note 4), certain governance modes were delegated to the things, which then did their governing job without being recognized as such. Since actual nano 'things' are still (mostly) in the future, such delegation is not possible. But there are expectations, and these can solidify into a forceful societal agenda that will govern strategic choices. One might call this 'delegation to the future', and one can definitely see such 'delegations' occurring in the domain of nanotechnology.

De facto risk governance in the domain of nanotechnology

Before tracing the emergence of a risk governance agenda, it is important to note that 'nanotechnology' is an umbrella term, covering quite different scientific and technological developments that are similar only in that they work at the nano scale. In policy making, and to some extent in media coverage and public perception, it is the umbrella term that is used, so that differences are black-boxed even where they would be relevant.[6] In the risk debates, the reference is often just to nanotechnology, while the actual concerns, as well as present studies, are about nanoparticles (and engineered free nanoparticles at that). A reconstruction of the emergence of a forceful agenda will have to take this into account and maybe explain the focus on nanoparticles.

For a reconstruction of the evolving risk governance debate and the resulting *de facto* agenda, I build on a study by Van Amerom and Rip, based on a comprehensive study of documents (up till 2006) and on interviews and participant observation in relevant meetings, and partly reported in Rip and Van Amerom (2010). This study focussed on societal and *de facto* agenda-building as the key phenomenon rather than the traditional focus in agenda-building studies on one single arena and what happens inside that arena. Societal agenda-building is a multi-arena process, without there being a clear authority deciding on the agenda. Kingdon (1984) provided the starting point for this analysis with his discussion of policy entrepreneurs and their skills, their networks, as well as how they can act on policy windows and other opportunities to forge a new, or change the existing, agenda. This converges with a point made in sociology: 'Arenas and fora, and the various issues discussed and addressed there, [which] thus involve . . . political activity but not necessarily legislative bodies and courts of law' (Strauss, 1978: 124). Such (always partial) entanglements can become locked-in into a forceful agenda, and then lead to path dependencies (Rip et al., 2007).

Figure 7.1, reproduced from Rip and Van Amerom (2010), maps the emerging paths in the evolution of the debate and activities and strategies. Time is on the horizontal axis, and the visualization of the developments begins with promises about application of nanoparticles as voiced around 2000 and taken up by researchers and firms. The vertical axis comprises ongoing practices of production and use of nanoparticles, then meso-level activities of collective organizations (and of research and regulation), and macro-level societal debate. While there was already a general idea about the promises of nanotechnology, linked to the establishment of the US National Nanotechnology Initiative in 2000, and some concerns were voiced based on speculative scenarios about run-away nano-robots, macro-level debate proper only started with a Canadian-based NGO (ETC group) issuing an early warning about risks of nanoparticles and calling for a moratorium on their development. This was resisted by nano-enactors, but was listened to during 2003 by governance actors such as the European Parliament and the UK government (see for details Rip and Van Amerom, 2010).

For my analysis of *de facto* governance, I highlight two points visible in the further developments. The first is that arenas overlap, and that actors, in practice, are

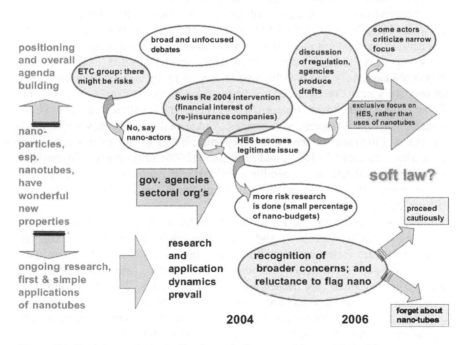

Figure 7.1 Evolving paths in the 'landscape' of nanoparticles and their risk

not limited to their formal roles. Government actors with regulatory responsibility (especially those who are pro-active) attend meetings and generally take part in a variety of arenas where informal societal agendas are built. Industrial actors mingle with other kinds of actors, especially if a somewhat neutral space is provided. After the intervention in the debate, in 2004, by one of the main re-insurance companies Swiss Re, with a report arguing that carbon nanotubes might create risks similar to those of asbestos, risks of nanoparticles became a legitimate topic. A subsequent meeting organized by Swiss Re and the International Risk Governance Council (IRGC) in Zürich in July 2006 was an occasion for informal interactions.[7] The meeting discussed a (potentially authoritative) report on risk governance of nanotechnology authored by Renn (a risk and public deliberation scholar) and Roco (of the US National Nanotechnology Initiative). Governmental and industry actors from across the world attended, as well as NGOs and scientists and scholars studying nanotechnology in society. Dedicated workshops and mingling in the corridors allowed interaction and the recognition of positions of other actors (and thus some learning). What was very visible was the recurrent anticipation of public reaction (in co-evolutionary terms, this can be described as possibly leading to selection before-the-fact). Clearly, the traditional distinction between formal agenda-building by authoritative (policy) actors and informal societal *de facto* agenda-building becomes blurred. While one can nonetheless recognize, with Shibuya (1996), that for a risk issue to rise on the formal agenda it needs to be taken up in both formal and informal agenda-building processes, articulation and prioritization processes are clearly not separate.

The second point is how actual soft law, as with the recent voluntary reporting schemes of UK Defra and US EPA (cf. Kearnes and Rip, 2009), is not just a matter of a new government initiative. It is prepared through actors moving in new directions. As is visualized in Figure 7.1, such actors can be firms that realize they need to proceed cautiously and possibly assure credibility by being more transparent. Or regulatory actors recognizing that there are openings for regulatory action but do not know exactly how to proceed. It is the combination of the two that creates a situation where soft law can be envisaged. And even then, there may not be much receptivity. Firms are reluctant to start reporting if they do not know what will be done with such data.[8] As Djelic and Andersson (2006: 378) note for transnational governance, 'Soft rules are generally associated with complex procedures of self-presentation, self-reporting and self-monitoring', and may thus lead to more organization rather than less.

Interestingly, some companies do take the initiative. Codes of conduct are formulated (see further section 3), and a Risk Framework for Nanotechnology has been put forward by the unusual alliance of a big chemical firm (DuPont) and a non-profit group (Environmental Defense). The alliance was announced in June 2005.[9] Their eventual risk framework, published in June 2007 following wide consultation, is a substantial contribution, even if the alliance has come in for criticism from (other) NGOs and trade unions.[10] In the Executive Summary (7), the authors actually note the link with government regulation:

We believe that the adoption of this Framework can promote responsible development of nanotechnology products, facilitate public acceptance, and support the development of a practical model for reasonable governmental policy on nanotechnology safety.

In other words, actors now contribute to evolving governance arrangements in a reflexive manner.

The implications of this discussion of risk governance agenda-building are two-fold. First, that risk assessment in the real world and risk management and regulation are part of larger dynamics, are shaped by it, and their effects (their 'success') are partly determined by these broader dynamics. One reason for the dominance of broader dynamics is the uncertainty about toxicity and exposure of nanotechnology materials (cf. Bowman and Hodge, 2006; Dorbeck-Jung, 2007). The point is general, however. The fate of risk assessments (i.e. their uptake and their 'translation') is not determined by their own 'internal' quality, but by their evolving contexts, which are influenced by other/earlier risk assessments and debates, in this case on genetically modified organisms, and earlier, on nuclear energy. Interestingly, in both earlier cases a storyline of the escape of modified micro-organisms and a run-away nuclear reactor occurred, horror stories which returned in the shape of nano-robots getting out of hand. Similarly, regulation is only one element in a range of governance activities and arrangements, which all operate at the interface between nanotechnology and policy/society and add up to a governance 'landscape' (Kearnes and Rip, 2009). A key element of this landscape, and in a sense a precondition for regulation, is the process by which a *de facto* risk agenda emerged and shaped responses and interactions.

Second, the actions and reactions that build-up to a socio-technical agenda, which solidifies and then shapes further actions and choices, create patchwork governance arrangements rather than a coherent system. This is clear in the way voluntary reporting and other soft law approaches are progressing (haltingly), as well as in the potential uptake of the Dupont & Environmental Defense Risk Framework and the critical reactions to it. Such patchwork arrangements may well allow nanotechnological innovations to continue, and in that sense be seen as productive. They may turn out to be inadequate, however, when something untoward happens, for example when an unusual toxic effect surfaces. The only politically viable response then is to clamp down on nanotechnology in general and introduce harsh precautionary measures.[11] This is not an argument against patchwork governance arrangements, but an indication of inevitable trade-offs.

Discourse and practice of responsible development of nanotechnology

Whereas the reference to risk, and thus to possible regulation, created some coherence in the evolving patchwork, there is also more open-ended *de facto* governance occurring around nanotechnology, linked to phrases like 'responsible development' and 'responsible innovation', and in the US, 'responsible steward-ship'.[12] The implications are rarely spelled out systematically, but the thrust can be captured in this quote from the US National Research Council:

> Responsible development of nanotechnology can be characterized as the bal-ancing of efforts to maximize the technology's positive contributions and minimize its negative consequences. Thus, responsible development involves an examination both of applications and of potential implications. It implies a commitment to develop and use technology to help meet the most pressing human and societal needs, while making every reasonable effort to anticipate and mitigate adverse implications or unintended consequences.
>
> (National Research Council, 2006: 73)

Clearly, further development of nanotechnology is the main goal, but openings are created for considering 'adverse implications or unintended consequences' and perhaps for doing something about them. This may invite nano-promoters to consider broader issues at an early stage, and allow other actors to raise questions about the present direction of development.

European Commission documents on nanotechnology often refer to respon-sible innovation and, recently, a further step was taken by preparing and publish-ing a code of conduct for nanoscience and nanotechnology (N&N) research.[13] The restriction of the code to 'research' was necessary, because of the limited remit of the European Commission in this respect, but the code is broader, and refers also to public understanding and the importance of precaution. There are explicit links to governance: the guidelines 'are meant to give guidance on how to achieve good governance'; as the Commission further specifies,

> Good governance of N&N research should take into account the need and desire of all stakeholders to be aware of the specific challenges and opportu-nities raised by N&N. A general culture of responsibility should be created in view of challenges and opportunities that may be raised in the future and that we cannot at present foresee.

A 'general culture of responsibility' cannot be created by the European Commis-sion, of course, but they clearly see themselves as contributing to such *de facto* governance.

US and European government actors are not alone in pushing 'responsible development'. There are now also codes of conduct (specifically for nanotechnol-ogy) formulated by firms like BASF addressing the corporation's responsibili-ties to 'our employees, customers, suppliers and society but also towards future generations',[14] and similar statements, for example by Degussa (now Evonik).[15] Recently, the Swiss retail industry went through the exercise of formulating a code.[16] Then there is also the recent initiative toward a 'Responsible Nanotech-nologies Code' led by a group consisting of the UK Royal Society, an NGO (Insight Investment), the Nanotechnology Industries Association, and supported by a network organized by the UK Department of Trade and Industry.[17] The proposed code goes much further than merely the safe handling of nanotechnology, but it is not clear if and how it will be taken up. Negotiations about a final text that can be made public are still ongoing.

There has been criticism of codes of conduct as being bland (though not all of them are) and as not specifying sanctions. Even so, they create openings for accountability. This does imply that it depends on others willing to call nano-enactors to account for whether the codes will have a real effect. While the discourse of responsible development will have implications for practices, broader issues referred to in the discourse will often be pushed to the background. The actual operationalization of 'responsible innovation' tends to focus on risk issues, transparency, and some public dialogue; and in the case of industry, also as a responsibility for the safe handling of nano-production and nano-products.

While the recent emergence of Codes of Conduct (actual and proposed) already indicates distributed governance, they should be seen as the tip of an iceberg of anticipatory and reflexive actions and interactions that fill the gap between further the development of nanotechnologies and the actual and possible responses by society. There is a plethora of activities and gatherings in the nano-world with governance elements and/or implications, with explicit or implicit reference to responsible development. One can see them as emerging practices of discussion, deliberation, negotiation, and participation.

In some cases, responsible development is a secondary effect. Definitions and standards for nanotechnology are of immediate importance for co-ordination among firms, but they will also be used to indicate the scope of regulation and of soft law like voluntary reporting. The International Standards Organization (ISO) has established working parties, and its standards are sometimes called 'soft governance'.[18] They are voluntary but recognized as important and authoritative because of the process leading to them (expert working parties and wide consultation). Actors/stakeholders refer all the time to ISO standards and the working parties because they expect them to resolve uncertainty. The OECD has also become involved, and also looks at risks and public engagement; its working parties have a certain status and are expected to come up with authoritative conclusions. UNESCO has invested in a report on ethics of nanotechnology (UNESCO, 2006). Further, dedicated groups or associations have been established, for example the International Council On Nanotechnology (ICON) has collected many stakeholders (but almost no NGOs). A web of activities and interactions results, and actors in the nano-world can refer to it to show that responsible development is being taken seriously.

Policy actors and nanotechnology spokespersons from industry try to keep abreast of what is happening and thus monitor the evolution of the discourse and the positioning of the various actors and groups. The director of the nanotechnology R&D programme in the European Commission's 6th and 7th Framework Programmes (until September 2008), Renzo Tomellini, often showed a slide with an overview (Figure 7.2). The fact that he shows it, and updates it, is just as important as the content of the slide.[19] The link with responsible development is clear

Main International Fora and Initiatives on Nanotechnology

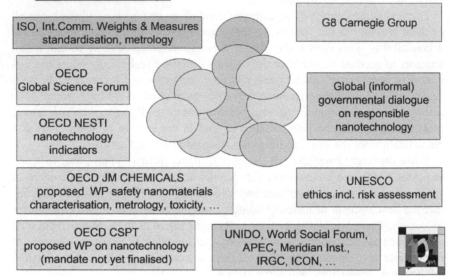

ISO, Int.Comm. Weights & Measures standardisation, metrology	G8 Carnegie Group
OECD Global Science Forum	Global (informal) governmental dialogue on responsible nanotechnology
OECD NESTI nanotechnology indicators	
OECD JM CHEMICALS proposed WP safety nanomaterials characterisation, metrology, toxicity, …	UNESCO ethics incl. risk assessment
OECD CSPT proposed WP on nanotechnology (mandate not yet finalised)	UNIDO, World Social Forum, APEC, Meridian Inst., IRGC, ICON, …

Figure 7.2 Tomellini's overview of activities in the nano-world (version summer 2007)

in his mind. When (in meetings in 2007) presenting data on public opinion about nanotechnology, he was happy to note that the European public is more positive than the North-American public. 'So we have done a good job. But this trust in us also creates a responsibility to make sure that nanotechnology is developed in the right way'.

Another action that is particularly interesting for what it might do (rather than what it does at the moment) is the International Dialogue on Responsible Development of Nanoscience and Nanotechnology, set in motion by Mike Roco (US National Nanotechnology Initiative) and Renzo Tomellini (European Commission's Nanotechnology Program) in 2004, and perhaps now gathering a momentum of its own. The idea was and is to have informal interactions between government officials and other actors in the nano-world, with reference to responsible development as one reason why coordination is important. After the first meeting in Alexandria, Virginia (Meridian Institute, 2004), there was a delay because of political difficulties, but then meetings were held in Tokyo (2006) and Brussels (2008), with a next meeting planned in Russia. Such meetings offer reporting on developments, including ethical and social aspects (ELSA) and experiences with public dialogues, but are also space for interaction. Their advantage is that they can be inclusive: there is no official mandate or link to an authority, so no actual or symbolic barriers to participation.

The key point to draw out of this mapping of activities is how a variety of actors begin discussing responsible development of nanotechnology, refer to it, and develop relevant activities, and how the discourse shows some convergence. This may be the beginning of a shift in governance, driven by prudence and some good intentions, as well as the need to maintain legitimacy. Policy actors may be involved, but often interactively rather than stipulating a governance arrangement, and they build on receptivity to the discourse on responsible development that is clearly out there. Thus, there is some *de facto* governance of nanotechnologies. A subsequent question then is whether this is specific to the nature and situation of nanotechnologies, or whether it reflects a general shift in governance, in the direction of reflexive modernization (Beck, 1992; Beck et al., 2003). The latter will be the case, definitely, but nanotechnologies offer a 'lead' domain where the shift is visible, and is, in a sense, experimented with.

An overarching pattern?

There are more activities and emerging structures to be mapped, not necessarily specific to nanotechnologies, but taken up in earnest there. Upstream public engagement, including citizen conferences after the Danish model, is one example (Kearnes et al., 2006), the interest in having ELSA (Ethical, Legal, and Social Aspects) included in the big nano-research funding programmes in the US and Europe another.[20] Together with codes of conduct of varying status, and emerging soft law with a precautionary flavour, these fill the gap between promotion and control of nanotechnologies. Thus, the traditional governance arrangement of new technologies is shifting. The new activities and interactions are expected, and this will orient (enable and constrain) strategies, actions, and interactions, and will be seen as legitimate.

We see an emerging overarching pattern, but its strength and eventual shape are still unclear. A key component of the pattern is anticipatory governance, and in particular, the consideration, at an early stage, of eventual societal embedding, including an acceptance of some responsibility. Will what is still mainly discourse become a practice, and a practice of *de facto* governance at that? To explore this question further, I will mobilize insights from STS and economics of innovation, because there is a structural similarity with patterns that have emerged in technological development *per se*.

Evolutionary economics and sociology of technological development have identified (and theorized) so-called 'trajectories' of development at two levels. There are paths of successive specific designs and products, related to what Dosi (1982) called the 'technological paradigm', which shapes expectations about productive directions of development and thus defines requirements. Nelson and Winter (1977: 56) refer to technicians' beliefs about what is feasible or at least worth attempting.

> When such beliefs and corresponding design and development practices are entrenched, one can speak of a 'technological regime' determining

technological decisions. An example is how the advent of the DC3 aircraft in the 1930's defined a particular technological regime: metal skin, low wing, piston powered planes. Engineers had some strong notions regarding the potential of this regime. For more than two decades innovation in aircraft design essentially involved better exploitation of this potential; improving the engines, enlarging the planes, making them more efficient.

Then, there are broad design heuristics or guidelines like mechanization (since the nineteenth century), and automation (from the 1950s onwards), what Nelson and Winter (1977) called a 'natural trajectory' and I will call a second-order trajectory to bring out the relation to technology-specific and thus first-order trajectories.

In specific developments in biotechnology, genomics, stem cells, and nano-technologies, one can find first-order trajectories, some still emerging, others already established. For micro- and nanotechnologies, there is a second-order trajectory of miniaturization. Such overall requirements on the development of new technologies constitute a governance pattern. My additional point now is that the requirements need not be limited to those that specify technical performance.[21]

At present, for all newly emerging technologies, one sees attempts to include societal aspects and to anticipate embedding in society. Already due to credibility pressures, such anticipation functions as a soft limit on specific developments. Thus, one can hypothesize that a further second-order trajectory may emerge, which could be labelled 'working towards the adequate societal embedding' of technology. There is no guarantee that it will indeed become a trajectory of tech-nological development even if policy actors are keen on it, as is visible in the discourse of 'responsible innovation'. Just as with mechanization and automation, the second-order trajectory of 'working towards adequate societal embedding' need not be taken up in all concrete developments for it to function as a gover-nance arrangement. But there must be sufficient actual uptake, and broad refer-ence to anticipation on societal embedding to make it a second-order trajectory. This emerging second-order trajectory is not specific for nanotechnology, but it is in this area that the indicators are strongest – it is the lead technological domain for the trajectory.

Part of such a second-order trajectory seems to be in place already: the inclu-sion of EHS (Environmental, Health, and Safety) aspects in technological devel-opments at an early stage. This actually builds on what one could call an earlier internalization of requirements from the selection environment: the chemical industry's Responsible Care programme in the 1990s (King and Lenox, 2000). It is significant that firms presenting a code of conduct (or similar) for nanotechnol-ogy are chemical companies. The focus on EHS may create a lock-in: in nano-technology, as well as for other new technologies like GMO, adequate societal embedding will quickly be reduced to EHS aspects. Some actors, however, like Degussa (a chemical company) do emphasize the importance of responsibility and dialogue, and attempt to interact with new actors like critical NGOs. It is uncertain whether such operationalizations of broader anticipation will be suc-cessfully internalized.

An indirect indicator of internalization is that funding agencies start creating special programmes on ELSA and societally responsible innovation.[22] In a sense, funding agencies are 'third parties': they do not develop (nano)technology themselves, but influence developments through their actions.[23] Capital providers like banks and pension funds (and perhaps even venture capitalists looking further than an immediate return on investment) might play such a third-party role as well, where they were to introduce requirements of responsible development. When funding agencies and other sponsors of research and development actually require anticipation on adequate societal embedding, nano-enactors have to develop relevant competencies, and this will contribute to a solidification of the arrangement.

In conclusion

Broadening the notion of governance has enabled me to discuss *de facto* governance, both in general and how it occurs in the domain of nanotechnology. There were attempts at intentional governance, addressing uncertainties around an emerging technology, but their fate had to be understood against the backdrop of evolving *de facto* governance. The importance of societal agenda building through the interactions of actors and their strategies was clear. For risk governance, our understanding of such emerging patterns might be used to create scenarios of possible futures, to improve anticipatory governance, or at least to increase reflexivity. The discourse of responsible development also showed a mixture of efforts towards intentional governing (as in the case of the European Commission's Code of Conduct for N&N research), actors' initiatives, and emerging patterns in a web of interactions creating orderings in the world of nanotechnologies. These are general features of *de facto* governance.

What was striking in the emerging *de facto* governance of nanotechnologies was the role of anticipation. Actors anticipate both possible futures as well as the reactions of other actors. This is more than prudence: newly emerging technologies like nanotechnologies create openings (opportunities as well as some concerns) that are uncertain by definition. The future becomes a reference point, even if it is unknown. One can speak of the 'shadow of the future', in the same vein as Scharpf (1997) talks of the 'shadow of hierarchy'. Scharpf uses the metaphor of 'shadows' more widely than simply in reference to hierarchy, e.g. 'in the shadow' of the majority vote (191) or 'in the shadow' of a statute (202), without being explicit about the actual mechanisms and dynamics at play (other than his reference to anticipated reactions when discussing decision making in bureaucracies (200)). For Scharpf, a key notion is 'authority structure' and how this can be referred to and thus have effects in an indirect way. Thus, for new and emerging technologies like nanotechnologies, 'the future', when articulated in more or less forceful societal agendas and expectations about responsible development, functions as an authority structure and thus casts its shadow on choices and actions in the here and now.

De facto governance is distributed, almost by definition, and cannot be easily shaped from a central point. This introduces ambivalence in the role of governance actors like governments. They have to give up on the assurance of governability as lots of things are outside their power and influence. At the same time, there is *de facto* governability: social orders are there and are (somewhat) effective, even if the direction may not be ideal. (compare Rip and Groen 2001) Moreover, one can see intentional and *de facto* governance co-evolving, and governance actors might then see their role as one of modulating this co-evolution (Rip, 2006).[24] Where UK Defra and US EPA opted for voluntary reporting, they hoped to draw on a sense of responsibility of firms, but clearly they were too early to effectively modulate: the evolution on the other side had not progressed sufficiently (the situation might be different in continental Europe). The International Dialogue discussed in section 3 began instead from the other side: while government actors were involved, they do not govern, but create instead a space in which *de facto* governance might be stimulated. Modulation of co-evolution also occurs in the public dialogues, in the creation of voluntary codes and responses by civil-society actors, and in technology assessment interactions visible in the world of nanotech-nologies. It is explicitly mentioned in the recent calls for 'midstream modulation' (Fisher et al., 2006) and in midstream public engagement (Joly and Rip, 2007). The work of governing is distributed, and may become partially internalized when regimes stabilize. The possibility of a second-order trajectory where working towards adequate societal embedding as a requirement of ongoing technological developments is a particular, and particularly interesting, case.

Even without a fully fledged second-order trajectory, nano-enactors already take initiatives by themselves or stimulated by third parties. This works out dif-ferently in the various domains under the umbrella term 'nanotechnology'. For new materials, chemical companies have relevant competencies because of the earlier (and continuing) Responsible Care Programme, and they feel credibility pressures. It is in this sector that firms have come up with nanotechnology codes of conduct. Micro-electronics firms, on the other hand, who do a lot of work at the nano-scale, are far removed from end users (even if they try to create some vis-ibility, as with labels 'Intel inside' on laptops). There are discussions, for example about RFID and about ambient intelligent systems enabled by nanotechnology, but there are other firms take the lead.[25] Big incumbents have the resources to be pro-active, but do not always rise to the occasion. In bio-nanotechnology, the third main domain of nanotechnology, the big pharmaceutical companies are interested, but tend to wait for small firms to come up with nano-enabled innovations like diagnostic devices and drug delivery. For small firms, their first concern is sur-vival, and broader anticipation is a luxury. Third parties like insurance companies may be able to (sometimes inadvertently) modulate productively. Other input in the *de facto* governance of nanotechnologies is likely to come from interactions across the product-value chains. The first signs of this are linked to health and environmental issues.

Thus, *de facto* governance is not blind. It is shot through with attempts at shap-ing, and by their residues, somewhat stabilized regimes around nanotechnologies, and newly emerging technologies more generally. The question of the quality of such governance, e.g. governability, legitimacy, and the directions that are

pushed, is important but remains difficult to consider since no one actor is specifically responsible. However, as soon as regimes and second-order trajectories
appear, these offer entrance points for critical evaluation and perhaps attempts at
changing them – by modulation.

Notes

1 This point has been made in implementation studies, starting with Pressmann and Wildavsky (1984) and becoming almost a movement (of the 'bottom-uppers') in the 1980s
(see Hanf and Toonen, 1985). There is a tendency, however, to push to this background
in actual policy making and implementation, because policy makers must show that it
is they who are making a difference.

2 Cited in Voß (2007: 34).

3 At an early stage, however, it will be unclear what sort of performance and side effects
might be realized. This adds up to a dilemma of knowledge and control (Collingridge,
1980), which has become one motivation to do technology assessment at an early
stage.

4 An example the 'given' character of sociotechnical governance arrangements, often
quoted in the STS literature, are the overpasses on Long Island, which continue to
'govern' what is possible and what is not possible even after Robert Moses' original
intentions became irrelevant (Winner, 1980). Their designer, New York city architect
Robert Moses, created them to keep New York's black and poor whites (who had to
use buses at the time, the 1920s and 1930s) away from the beaches and parks he had
created on Long Island. He tried to create a material constitution for his preferred social
order, and while it may have worked for a time, this particular constraint on behaviour
has become irrelevant now that every American can use a motor car.

5 This is an Actor-Network Theory notion, cf. how Latour (1991), for similar reasons,
speaks of a 'Parliament of Things'. See also Verbeek (2006) on the morality of artefacts.

6 There is a *de facto* governance element involved in such processes: some terms become
forceful exactly because they remain blackboxed.

7 The IRGC is a private not-for-profit foundation based in Geneva, 'to support governments, industry, NGOs and other organizations in their efforts to understand and deal
with major and global risks facing society and to foster public confidence in risk governance' (cited in Renn and Roco, 2006: 5). A conference report is available from Swiss
Re Center for Global Dialogue.

8 By July 2008, only nine companies had registered with the Defra scheme and EPA
had received four submissions under the basic programme (and commitments from
12 more companies), whilst no company has agreed to participate in the in-depth programme. Interestingly, some branch organizations, recognizing the importance of the
scheme for the credibility of the nanotechnology sector, tried to push their members to
participate (see Kearnes and Rip, 2009).

9 The two partners had a sense of the historical importance of their attempt when they
announced it in an article in *The Wall Street Journal*, 14 June 2005, under the title
'Let's Get NanoTech Right'. This echoes an earlier claim about how to handle nanotechnology: 'Let's get it right the first time!' (Cf. Roco and Bainbridge, 2001)

10 Partly in response, an 'international coalition of [seven] consumer, public health, environmental, labor and civil society organizations spanning six continents called for
strong, comprehensive oversight' of nanotechnology and nanomaterials. Their text has
a strong precautionary thrust; 'voluntary initiatives are not sufficient'. Quoted from the
press release, 1 August 2007 (www.nanowerk.com/news/newsid=2306.php).

11 This is actually one of the three scenarios developed by Douglas Robinson for a Constructive Technology Assessment workshop on responsible innovation in nanotechnology, December 2007. See, for the methodology, Robinson (2009).

12 See for example the proposal, in California, for a Nanotechnology and Advancement of New Opportunities (NANO) Act by Rep. Honda (D- San José): 'The NANO Act requires the development of a nanotechnology research strategy that establishes research priorities for the federal government and industry that will ensure the development and responsible stewardship of nanotechnology'. Compare www.house.gov/apps/list/press/ca15_honda/NanoAct2008.html

13 C(2008)424 final, 7 February 2008: Commission Recommendation on a code of conduct for responsible nanosciences and nanotechnologies research.

14 *Code of Conduct Nanotechnology.* http://corporate.basf.com/en/sustainability/dialog/politik/nanotechnologie/verhaltenskodex.htm?id=VuGbDBwx*bcp.ce. (accessed 4 March 2008).

15 Degussa's website on nanotechnology has an item to this extent on responsibility (www.degussa-nano.com/nano).

16 Interessengemeinschaft Detailhandel Schweiz. 2007: *Code of Conduct: Nanotechnologies.* www.cicds.ch/m/mandanten/190/download/CoC_Nanotechnologien_final_05_02_08_e.pdf.

17 See the May 2008 update at www.responsiblenanocode.org.

18 See presentation by Peter Hatto (ISO) at the International Dialogue meeting, Brussels, March 2008. (Tomellini and Giordani, 2008)

19 There is now an official version (without the clouds in the middle), published on Cordis. ftp://ftp.cordis.europa.eu/pub/nanotechnology/docs/a-interactions-global.pdf

20 The US National Nanotechnology Initiative has funded two big Centres for Nanotechnology in Society, and some smaller units. NanoNed (www.nanoned.nl).

21 While used in another context, the notion of meta-rules of the game (Djelic and Andersson, 2006: 385, 391) indicates a similar phenomenon.

22 In the Netherlands, such a programme has just started; it focuses on (a) advanced (emerging) technologies and (b) sociotechnical system transitions (www.nwo.nl/mvi). In Norway, the theme is ELSA of biotech, nanotech and neurotech (www.forskningsradet.no). In both cases, interaction between social science and humanities on the one hand and science and engineering on the other is an important requirement. In the UK, the Engineering and Physical Sciences Research Council (EPSRC) established a Societal Issues Panel in 2006 and experimented with dialogue. In the words of a participant observer, 'An emergent sociotechnical imaginary that takes "societal issues" not as an obstacle but as an active contributor to framing the work of the research councils' (Doubleday, 2008).

23 Analytically, the importance of such 'third parties' taking initiatives is that they can break through waiting games and other impasses that occur often in two-party games (Scharpf, 1997). An example of such a breakthrough is the intervention of re-insurance company Swiss Re in the risk debate (see section 2).

24 This is similar to what Andrew Dunsire (1996) has called 'collibration' – an intervention to shift the pre-existing balance between countervailing forces. Institutional arrangements that have the effect of strengthening one or weakening the other of these forces will require much less energy than institutions that would have to stop an unopposed force. (Taken from Scharpf, 1997: 182).

25 Firms like Philips and Siemens, who used to cover both sectors, have now divested their micro-electronics development and production.

References

Anderson, B., *Imagined Communities: Reflections on the Origin and Spread of Nationalism* (Revised Edition, London: Verso, 1991).

Barben, D., Fisher, E., Selin, C., and Guston, D.H., 'Anticipatory Governance of Nanotechnology: Foresight, Engagement, and Integration', in Edward J. Hackett et al. (eds.),

The Handbook of Science and Technology Studies (Third Edition, Cambridge, MA: MIT Press, 2007).

Beck, U., *Risk Society: Towards a New Modernity* (London: Sage Publications, 1992).

Beck, U., Bonss, W., and Lau, C., 'The Theory of Reflexive Modernization: Problematic, Hypotheses and Research Programme', 20 *Theory, Culture & Society* (2003), 1–33.

Bowman, D.M., and Hodge, G.A., 'Nanotechnology: Mapping the Wild Regulatory Frontier', 38(9) *Futures* (2006), 1060–1073.

Braithwaite, J., and Drahos, P., *Global Business Regulation* (Cambridge: Cambridge University Press, 2000).

Collingridge, D., *The Social Control of Technology* (London: Frances Pinter, 1980).

Djelic, M.-L., and Andersson, K.S. (eds.), *Transnational Governance: Institutional Dynamics of Regulation* (Cambridge: Cambridge University Press, 2006).

Dorbeck-Jung, B., 'What Can Prudent Public Regulators Learn From the United Kingdom Government's Nanotechnological Regulatory Activities?', 1 *Nanoethics* (2007), 257–270.

Dosi, G., 'Technological Paradigms and Technological Trajectories: A Suggested Interpretation of the Determinants and Directions of Technical Change', 11 *Research Policy* (1982), 147–162.

Doubleday, R., 'No Room for Doubt: Public Engagement, Science Policy and Democracy at the UK's Engineering and Physical Sciences Research Council', Paper presented at the meeting of the Society for Social Studies of Science and the European Association for the Study of Science and Technology, Rotterdam, 21–23 August 2008.

Dunsire, A., 'Tipping the Balance: Autopoiesis and Governance', 28 *Administration and Society* (1996), 299–334.

Fisher, E., Mahajan, R.L., and Mitcham, C., 'Midstream Modulation of Technology: Governance From Within', 26(6) *Bulletin of Science, Technology and Society* (2006), 485–496.

Hanf, K., and Toonen, T.A.J., *Policy Implementation in Federal and Unitary Systems: Questions of Analysis and Design* (Dordrecht: Martinus Nijhoff Publishers, 1985).

Joly, P.-B., and Rip, A., 'A Timely Harvest', 450 *Nature* (8 November 2007), 308.

Kearnes, M.B., Macnaghten, M., and Wilsdon, J., *Governing at the Nanoscale: People, Policies and Emerging Technologies* (London: Demos, 2006).

Kearnes, M.B., and Rip, A., 'The Emerging Governance Landscape of Nanotechnology', in S. Gammel, A. Lösch, and A. Nordmann (eds.), *Jenseits von Regulierung: Zum politischen Umgang mit Nanotechnologie* (Berlin: Akademische Verlagsanstalt, 2009).

King, A.A., and Lenox, M.J., 'Industry Self-Regulation Without Sanctions: The Chemical Industry's Responsible Care Program', 43(4) *The Academy of Management Journal* (2000), 698–716.

Kingdon, J.W., *Agendas, Alternatives and Public Policies* (Boston: Little, Brown and Company, 1984).

Kooiman, J., *Governing as Governance* (London: Sage Publications, 2003).

Latour, B., *Nous n'avons jamais été modernes. Essai d'anthropologie symétrique* (Paris: La Découverte, 1991).

Meridian Institute and National Science Foundation, *Report: International Dialogue on Responsible Development of Nanotechnology* (Washington, DC: Meridian Institute, 2004). http://meridian-nano.org/Final_Report_Responsible_Nanotech_RD_040812.pdf.

Mintzberg, H., *The Rise and Fall of Strategic Planning* (London: Basic Books, 1994).

National Research Council, *A Matter of Size: Triennial Review of the National Nanotechnology Initiative* (Washington, DC: National Academies Press, 2006).

Nelson, R., and Winter, S., 'In Search of a Useful Theory of Innovation', 6 *Research Policy* (1977), 36–76.

Ostrom, E., *Governing the Commons: The Evolution of Institutions for Collective Action* (Cambridge: Cambridge University Press, 1990).

Pels, D., Hetherington, K., and Vandenberghe, F., 'The Status of the Object: Performances, Mediations, and Techniques', 19(5/6) *Theory, Culture & Society* (2002), 1–21.

Pressman, J.L., and Wildavsky, A., *Implementation: How Great Expectations in Washington Are Dashed in Oakland* (Third Edition, Berkeley: University of California Press, 1984).

Renn, O., and Roco, M., *Nantechnology Risk Governance* (Geneva: International Risk Governance Council, 2006).

Rip, A., Misa, Th., and Schot, J.W. (eds.), *Managing Technology in Society: The Approach of Constructive Technology Assessment* (London: Pinter Publishers, 1995).

Rip, A., and Groen, A., 'Many Visible Hands', in R. Coombs, K. Green, V. Walsh, and A. Richards (eds.), *Technology and the Market: Demands, Users and Innovation* (Cheltenham: Edward Elgar, 2001), pp. 12–37.

Rip, A., 'A Co-Evolutionary Approach to Reflexive Governance: And Its Ironies', in J.-P. Voß, D. Bauknecht, and R. Kemp (eds.), *Reflexive Governance for Sustainable Development* (Cheltenham: Edward Elgar, 2006), pp. 82–100.

Rip, A., Robinson, D.K.R., and te Kulve, H., 'Multi-Level Emergence and Stabilisation of Paths of Nanotechnology in Different Industries/Sectors', Paper prepared for International Workshop on Paths, Berlin, 17–18 September 2007.

Rip, A., and te Kulve, H., 'Constructive Technology Assessment and Sociotechnical Scenarios', in E. Fisher, C. Selin, and J.M. Wetmore (eds.), *The Yearbook of Nanotechnology in Society, Volume I: Presenting Futures* (Berlin: Springer, 2008), pp. 49–70.

Rip, A., and van Amerom, M., 'Emerging de facto Agendas Around Nanotechnology: Two Cases Full of Contingencies, Lock-Outs and Lock-Ins', in S. Maasen, M. Kaiser, M. Kurath, and C. Rehmann-Sutter (eds.), *Deliberating Future Technologies: Identity, Ethics, and Governance of Nanotechnology* (Berlin: Springer, 2010).

Robinson, D.K.R, 'Complexity Scenarios for Emerging Techno-Science', *Technological Forecasting and Social Change*, 76(2009), 1222–1239.

Roco, M.C., and Bainbridge, W., *Societal Implications of Nanoscience and Nanotechnologies* (Boston: Kluwer Academic Publishers, 2001). Available at www.wtec.org/loyola/nano/NSET.Societal.Implications/

Scharpf, F.W., *Games Real People Play: Actor-Centred Institutionalism in Policy Research* (Boulder, CO: Westview Press, 1997).

Schot, J.W., Lintsen, H.W., Rip, A., and Albert de la Bruhèze, A.A. (eds.), *Techniek in Nederland in de Twintigste Eeuw. VII. Techniek en Modernisering. Balans van de Twintigste Eeuw* (Zuthphen: Walburg Pers, 2003).

Shibuya, E., 1996, '"Roaring Mice Against the Tide": The South Pacific Islands and Agenda-Building on Global Warming', 69 *Pacific Affairs* (1996).

Strauss, A., 'A Social World Perspective', 1 *Studies in Symbolic Interaction* (1978), 119–128.

Swiss Re, *Nanotechnology: Small Matter, Many Unknowns* (Zürich: Swiss Re, 2004).

Tomellini, R., and Giordani, J., *Report: Third International Dialogue on Responsible Research and Development of Nanotechnology* (Brussels: European Commission, 2008) ftp://ftp.cordis.europa.eu/pub/nanotechnology/docs/report_3006.pdf.

UNESCO, 'Division of Ethics of Science and Technology', *The Ethics and Politics of Nanotechnology* (Paris: UNESCO, 2006).

Van Kersbergen, K., and van Waarden, F., 'Governance as a Bridge Between Disciplines: Cross-Disciplinary Inspiration Regarding Shifts in Governance and Problems of Governability, Accountability and Legitimacy', 43 *European Journal of Political Research* (2004), 143–171.

Verbeek, P.-P., 'Materializing Morality: Design Ethics and Technological Mediation', 31(3) *Science, Technology & Human Values* (2006), 361–380.

Voß, J-P., Bauknegt, D., and Kemp, R. (eds), *Reflexive Governance for Sustainable Development* (Cheltenham: Edward Elgar, 2006).

Voß, J.-P., *Designs on Governance: Development of Policy Instruments and Dynamics in Governance*. PhD thesis, Enschede: University of Twente, defended 18 October 2007.

Winner, Langdon, 'Do Artifacts Have Politics?', 109(1) *Daedalus* (1980), 121–136.

8 Constructive technology assessment and the methodology of insertion

Arie Rip and Douglas K.R. Robinson

Published as

Constructive Technology Assessment and the Methodology of Insertion, in Neelke Doorn, Ibo van de Poel, Daan Schuurbiers, and Michael E. Gorman (eds), *Early Engagement and New Technologies. Opening Up the Laboratory* (Dordrecht: Springer Science + Business Media, 2014, pp. 37–53.)

Introduction

The two key elements of Constructive Technology Assessment (CTA), broadening technology development by including more aspects and involving more actors, and doing so on the basis of an understanding of the dynamics of technology development and its embedding in society, were identified in the mid/late 1980s in the Netherlands (Schot and Rip 1997). It was part of a larger perspective, laid down in the government's Policy Memorandum on Integration of Science and Technology in Society (Ministerie van Onderwijs en Wetenschappen 1984). On the basis of the Policy Memorandum, a Netherlands Organization for Technology Assessment (now Rathenau Institute) was established in 1986. One of its projects was to develop the approach of Constructive Technology Assessment (Daey Ouwens et al. 1987). In the Ministry of Education and Sciences and in the Netherlands Organization for Technology Assessment, perspectives and expertise from Science, Technology, and Society studies played an important role. The further development of CTA occurred in STS studies, linked to evolutionary economics of technological change (Rip et al. 1995), and in the evaluation of attempts to broaden technology development, as in social experiments with electric vehicles (Hoogma 2000; Hoogma et al. 2002). The CTA approach was taken up in studies in Canada, the UK, Australia, Denmark, and Sweden. And it was positioned as part of an overall move towards more reflexive co-evolution of science, technology, and society (Rip 2002).

Newly emerging technologies like nanotechnology, with their promises but also raising concerns about possible negative impacts, are a challenge for the CTA approach because the envisioned broadening of technology development must now be about possible future developments rather than current practices. Such a challenge had been recognized before, and could then be addressed systematically from the early 2000s onwards when the Dutch national R&D programme NanoNed, on nanoscience and nanotechnology, wanted to have a Technology Assessment (TA) component, and made funding available for PhD students and postdocs. The findings of this TA NanoNed programme are the basis for this chapter, located in the larger picture of reflexive co-evolution of science, technology, and society.

CTA is a 'soft' intervention, and studies and reports are an input, not the main result. For emerging technologies, two key components of a CTA activity are (1) the building of sociotechnical scenarios of possible technological developments and the vicissitudes of their embedding in society (based on extensive document study and field work); and (2) the organizing and orchestration of workshops with a broad variety of stakeholders. The scenarios help to structure the discussion in the workshops (Robinson 2010) and stimulate learning about possible strategies (Parandian 2012). Therefore, it is important to have scenarios of high quality and relevance, and which can be seen as legitimate by workshop participants.

Compared with other approaches as discussed in this volume, CTA activities take into account what happens on a variety of 'work floors': research laboratories, conferences, workshops, agenda setting and planning meetings, roadmapping events, public debates anticipating issues related to technology developments. A corollary is that the CTA actor has to move about, observe, and actively circulate in locations where actors are shaping the emerging paths of nanotechnology and how it will become embedded in society. We will call this 'insertion' by the CTA actor; to emphasize it is not just a practical matter of collecting data, but also part of the methodology of CTA, combining diagnosis of dynamics and some soft intervention.

The enterprise of CTA: goals and practices

While explicit goals for CTA were specified already in the 1980s, the actual approaches were also shaped by opportunities and circumstances that arose following its inception. Based on the experiences, there was further articulation of goals. This section is an attempt to take stock, by looking at overall goals, how these are linked to more concrete objectives, particularly for the case of emerging technologies, and what sort of concrete activities and methodologies are now in place.

CTA sees itself as part of the overall undertaking of TA, starting in the late 1960s. The background of this undertaking can be formulated, in retrospect, as a 'philosophy' of TA (cf. Rip 2001a):

Reduce the (human) costs of learning by trial-and-error – which characterized much of our handling of technology in society –, and do so by anticipating future developments and their impacts, and by accommodating these insights in decision making and implementation.

This is not easy because early signalling may not get a hearing – particularly if it is early warning (cf. Harremoës 2001). And it is not limited to commissioned

TA studies. It is a societal learning process, in which many actors participate. Actually, over the years, TA has moved in the direction of societal debate and agenda-building, at least with Rathenau Institute and some other European TA offices (Delvenne 2011).

Within TA, some of the specifics of CTA derive from a diagnosis of how the handling of technology in society has evolved: the separation of 'promotion' and 'control' of technology in our societies, which emerged in the nineteenth century and are still with us (Rip et al. 1995). It is a heritage of the industrial revolution of the eighteenth and nineteenth centuries, where technology development became a separate activity, carried by engineers and located in firms and public or semi-public research institutes. Culturally, a mandate to do so emerged: new technologies could then be developed as such, because they could be positioned as contributing to the progress of society, and therefore be accepted, almost by definition. Institutionally, an indication of the separation between 'promotion' and 'control' is the division of labour between government ministries, some promoting the development of new technologies and innovation, while other ministries consider impacts and regulation. TA emerged within this regime of handling technology in society, and was institutionalized at the 'control' side of the division of labour. An important argument was (and is) the asymmetry between technology development actors and society at large, with the latter coming in at a late stage, and little information about the technology. The asymmetry is structural, but TA would offer information and considerations to the 'control' side, and reduce the asymmetry.[1] CTA wants to compensate for the asymmetry in TA approaches, by focussing on technology development.[2]

Building on this diagnosis, CTA aims to bridge the gap between innovation and the consideration of social aspects which inform attempts at 'control', and in doing so, broaden technology development and its embedding in society. It is 'constructive' TA because it aims to be part of the construction of new technologies and their embedding in society. This was the starting point of the enterprise of Constructive TA (Daey Ouwens et al. 1987). These aims can then be taken as objectives for the design and execution of CTA activities. They require analysis of dynamics of technology development and its embedding in society, and the ways it is influenced/shaped – insights which can be translated into leverage for change. They are input into the preparation for concrete CTA activities like 'bridging' workshops with stakeholders in a technology domain. They are also building blocks for a theory of CTA (Rip 1992).

The rationale for pursuing these objectives stems from larger goals and perspectives, as was clear in how we developed a diagnosis of what is the case now in handling technology in society, with the implication that it should be improved. By now, a number of overlapping goals have been put forward. Taking an evolutionary perspective, the division of labour between 'promotion' and 'control' of technology in society is part of how technology and society co-evolved. One can then take a step back, and consider ongoing co-evolution of science, technology, and society, and in particular, how it is becoming more reflexive, for example through technology policy, technology foresight, and technology assessment (Rip

2002). Thus, one can work towards improving reflexivity of the co-evolution, in various ways – this implies some modulation of the co-evolution. This qualifies as a background goal for CTA and is linked to learning (cf. also Grin and Van de Graaf 1996). It has been emphasized in the studies in the TA NanoNed programme (e.g. Robinson 2010; Parandian 2012). Then, constructive in CTA refers to its being part of the construction of increased reflexivity in science, technology, and society.[3]

Broadening technology development and increasing reflexivity serve a purpose. To be explicit about this, Schot and Rip (1997) emphasized an overall goal served by CTA, of a better technology in a better society. It is important to keep such a substantive goal visible, in general but also because the CTA objective of including more actors is often taken as advocating more participation, and thus refers to a goal of democratization of technological development (Genus 2006; cf. also Callon et al. 2001 for an intermediate position). Of course, no one has a monopoly on goals for CTA. The point is that recognition of a goal has implications for what are appropriate CTA activities. The activities we describe in this chapter are appropriate to the overlapping goals we have outlined, so it is inappropriate to criticize them as being insufficiently democratic.

Signs of change

An increase in reflexivity of co-evolution of science, technology, and society is visible in the recent policy discourse about responsible development of new technology, and responsible innovation. There are now some attempts to implement this, especially in the domain of nanotechnologies. One example is the Code of Conduct for Responsible Nanosciences and Nanotechnologies Research (European Commission 2008), which can now be referred to in the Member States of the European Union. There is overlap with CTA objectives, in the sense that responsible development is a way to bridge promotion and control, by internalizing control at the side of technology development. This can still keep a focus on promotion, when 'responsible' is only modifying 'development'. When 'responsible' is emphasized the development itself might be queried, up to the possibility of stopping it.[4]

Thus, there are signs that the institutional separation of technology development and attempts at control (because of projected societal impact) is being bridged. At least, there are pressures to bridge and various attempts at handling these pressures. Of course, there were such pressures before, as when TA was proposed and started to become institutionalized in the 1970s. What is new is that anticipation on societal impacts is now seen as being also a responsibility of technology developers (see also Guston and Sarewitz 2002).

While the dichotomies (innovation vs. responsible, technology developers vs. users) remain visible, there are interactions and mixed approaches, and the situation evolves further. The domain of nanoscience and nanotechnologies turns out to be a site for experimentation and learning – including controversy. There is widespread uncertainty about impacts and risks, while there are also proposals for regulation, and NGOs which advocate a precautionary approach. There is additional uncertainty about consumer and citizen reactions to new nanotechnology-enabled products and processes, and innovators can fear for barriers to public acceptance and possibly a public backlash if something would go wrong. All this is to be expected. What is new is that innovation actors are asked by societal actors to account for what they do. This will set articulation processes in motion.[5] When some stabilization occurs, there will be *de facto* governance, i.e. steering and shaping of action that has some legitimacy, even if there is no formal authoritative basis as in law and regulation (Rip 2010a). Up to a modification of the division of labour, with responsible innovation becoming the responsibility of innovation actors, in interaction with various societal actors.

The experimentation and mutual learning that occurs in and around nanotechnology is now taken up for other emerging technologies like synthetic biology and ambitious technological ventures like geo-engineering. Thus, one can take learning in sectors and in society as a further overall goal, and formulate stimulation of such learning as a broad objective for CTA.[6] For new technologies, the point has been made that responsibilities are distributed, just like technological development itself (Von Schomberg 2007). The simple contrast between technology developers and users is inapplicable then. Interaction and mutual learning become important to overcome mismatches and fragmentation, in innovation as well as in 'distributed responsible development'. New 'divisions of moral labour' have to be invented, and one can see various actors exploring (even if reluctantly) possibilities (Rip and Shelley-Egan 2010).

Transforming objectives into activities

In the move from objectives to concrete activities, particularly for doing CTA about new technologies, some further conceptualizations are introduced – in effect, more building blocks for a theory of CTA.

Our diagnosis of a gap between promotion and control of technology at the societal level, and as we phrased it in the TA NanoNed programme, the gap between innovation and ELSA in a sociotechnical domain or sector,[7] can be detailed further, to the level of interactions, using Garud and Ahlstrom (1997). They distinguish 'insiders' (i.e. developers/promoters) and 'outsiders' (i.e. users/regulators) and show that their evaluations of technology are structurally different because of this difference in position. They also consider situations where insiders and outsiders interact, to some extent, calling these situations 'bridging events'. One of the examples they study are hearings conducted by a regulatory agency like the US Food and Drug Administration.

Their terminology of insiders and outsiders captures one aspect of the positions with respect to technology development, but assumes these positions are given. However, a firm developing technology for new products or processes of its own, may also be a user of products supplied by another firm and then position itself as an outsider, e.g. requiring quality assurance. When Garud and Ahlstrom (1997) discuss the difference in perspective between insiders and outsiders, they

speak of 'enactment' and 'selection' cycles, respectively, in which the two function. 'Enactment', a term from symbolic interactionism, here refers to technology
developers and promoters working to realize their goal and vision, 'enacting' their
project. Thus, a functional terminology is possible, of 'enactors' who realize the
technology and identify with the project of doing so, and 'comparative selectors'
who can consider different options to select from and do formal or informal versions of cost-risk-benefit assessment (Rip 2006).[8] Garud and Ahlstrom show how
enactors focus on their projections (i.e. informal scenarios) for further development of the technology and its embedding in society, and thus see society as a constellation of possible barriers which have to be overcome. If questions are raised
about the technology, such an enactor perspective will immediately see them
as indications of potential barriers, even when the questions are mainly inquiry
rather than criticism. The response of the enactor then is to emphasize the promise
of the new technology – with the corollary that the commentators, if still reluctant,
are positioned as being against progress. If this happens in the public domain, it
will incite further, and possibly more critical, responses (Swierstra and Rip 2007).

One concrete implication of this diagnosis of the two positions and related perspectives is that CTA workshops must have 'enactors' as well as 'comparative
selectors' as participants, so as to function as bridging events, where participants
can (in Garud and Ahlstrom's felicitous phrase) probe each other's realities. With
the right mix of participants, what happens in these CTA workshops will reflect
dynamics in the wider world, so they will be like a micro-cosmos. The workshop is also a protected space, where participants have the opportunity to consider
alternatives and the possibility of modifying their strategies and eventual interactions in the real world without there being immediate repercussions.[9] Still, the
wider world has its own dynamics, and these are important for eventual uptake
and effect of the CTA exercise.[10]

There is a further implication, given that we decided to develop sociotechnical scenarios as an input into the CTA workshops. Scenarios speak to an enactor perspective, in their projection of further development of a new technology.
But we introduce twists, showing unexpected shifts (for enactors) and repercussions. Stakeholders representing comparative selectors, from potential users to
regulators and NGOs, will be present in the workshops. Thus, in the interactions,
different perspectives as visible in the scenarios will come alive because their
protagonists are present. This will work out well only if the scenarios reflect what
is at stake in the worlds of the participants, otherwise they will be disregarded as
irrelevant. At the same time, the scenarios must offer challenges to participants'
understanding of the situation. This is where social-science insights (from innovation studies, from STS, and more generally) will have to come in, to improve the
quality of mutual probing in the workshops.[11]

In general, analysis and diagnosis of developments are necessary steps to prepare a CTA exercise and orchestrate it productively. One has to know about the
forces at play in the technology domain and the evolving relationships (or lack of
relationships) between stakeholders. A key point for understanding what happens
as well as the eventual construction of scenarios is that 'entanglements' occur,

existing and emerging mutual dependencies which guide and thus limit interactions and strategic choices (Rip 2010b). This shapes the way new technologies (in our cases, nanotechnologies) will materialize. In other words, the future is predicated on these patterns and dynamics: an 'endogenous future' (Rip and te Kulve 2008; Robinson 2009). The scenarios develop the endogenous future into a number of possible futures, each starting with certain interventions and interactions and then exploring responses, repercussions, and eventual outcomes.

For example in the case of possible nanotechnology applications in food packaging, studied by te Kulve (2011), there is reluctance with the producers and retailers to invest in it because of uncertainty about consumer acceptance, combined with uncertainty about eventual regulation of the products. The mutual dependencies have the form of a waiting game (Parandian et al. 2012), and if nothing happens, the waiting game will continue (thus, an endogenous future). Given this diagnosis, one can imagine that interventions occur attempting to break through the waiting game. This was the starting point for the construction of three scenarios. In scenario 1, 'Only a little nano', collaborations between academic and industrial researchers are sought and supported, but that leads to niche applications only. The big promise of nanotechnology is backgrounded. In scenario 2, 'Regulation helps', the concerns about health and safety aspects cast a shadow over the developments, and small companies move away from working on nano-applications, also because regulation might be strict (and thus make product development expensive). The big incumbents welcome regulation because it reduces the uncertainties, and they proceed – cautiously. In scenario 3, 'Thresholds are passed', some institutional entrepreneurs recognizing the barriers set up a consortium for product development and persuade consumer organizations and risk research institutes to participate, arguing that this is a way for them to have some influence on the shape of future technology. This creates legitimacy and further support becomes available for strategic research topics like nano-enabled improvement of barrier properties of paper and plastic packaging. Pharmaceutical companies then become interested as well.

Choices to be made

As is clear from this example, in constructing scenarios choices must be made about what to focus on, and what not. These choices can be discussed in the workshop, and alternatives may be considered. In general, the need to make choices in setting up the CTA activity is a challenge (and a task) for the CTA analyst, especially for emerging technologies like nanotechnology which live on promises: which expectations are to be taken into account as more realistic and/or more important? What is seen as important also depends, of course, on the position from which such expectations are voiced, e.g. by an enactor or a comparative selector. The CTA analyst can build on her knowledge of the domain and its dynamics, including expectations and investments in the different worlds in which a new technology option is being developed and will be embedded. But the challenge remains.

Which future to focus on (for monitoring, for assessment)?

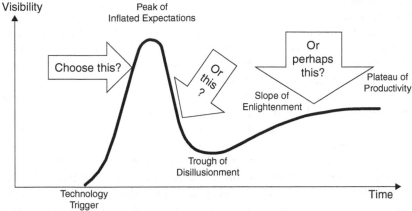

Figure 8.1 Gartner Group's hype-disappointment cycle[12]

The challenge can be brought out (even if in a somewhat simplified manner) by considering the hype-disappointment cycle, as introduced by Gartner Inc. Figure 8.1 shows the cycle, as well as different options for projecting a future state of the world. The realistic option (the eventual 'plateau of productivity') is also the most uncertain one, while relying on present promises may risk becoming victim of inflated expectations.

The risk is real, and not only in funding applications and other resource mobilization activities, where exaggerated promises are expected, and discounted. In discussions and activities exploring potential futures of a technology and its ethical and societal impacts, there is a tendency to go for the big impacts, so as to justify the effort to anticipate. It is all too easy then to extrapolate from current promises and end up in brave new worlds where human enhancement or interventionist ambient intelligence creates interesting ethical dilemmas. Nordmann and Rip (2009) have criticized such 'speculative' ethics of new technologies as disconnected from ongoing activities and the choices, ethical, and otherwise, that have to be made there. Our sociotechnical scenarios, building on endogenous futures, start from the other side. There is still speculation and imagination, of course, but it is not free-floating.[13]

For actors articulating their strategies the question of hype is a recurrent concern. Interaction with other relevant actors is important to reduce uncertainty, and in fact, the CTA workshops offer an opportunity to do so, and are appreciated for it. This was clear in the domain of Organic Large Area Electronics, studied by Parandian (2012). In one of his scenarios, he actually used the phenomenon of hype and disappointment, for nano-enabled RFID applications for security. This induced extended consideration of the value of government measures to realize the promises of a new technology.

So far, we have presented the CTA activities as doing a good job. And indeed they do, but some reflection is in order. CTA for new technologies aims to broaden design and development, at an early stage. Thus, it has an upstream bias: better outcomes result from doing better at an earlier stage. It is a bias, because it is the overall co-production process that leads to eventual outcomes, there is no determinism. But it is an unavoidable bias if one wants to address new technologies – which are by definition still at an early stage.

Upstream public engagement (in the UK and elsewhere) has the same bias, but in contrast with CTA it focuses on actors with little or no agency. They may well remain empty exercises, even if the views and discussions reported might be taken up by policy makers when they see fit. CTA addresses stakeholders, and does more than just solicit views from stakeholders. There are orchestrated bridging events, and there must be something at stake, for the participants and developments in the domain or sector. Looking back at the almost 20 CTA workshops we organized in the TA NanoNed program, we see that the less successful ones indeed suffered from there being little at stake (Robinson 2010).

A methodology of insertion

The aims of CTA to broaden technological design and development and make it more reflexive imply an action-orientation of CTA. CTA agents are change agents, but softly, through support and attempts at opening up, rather than pushing. If there is pushing, it is a push for more reflexivity (cf. Schot and Rip 1997). Theoretically and practically, this relates to the rationale of making the co-evolution of science, technology, and society more reflexive (so there will be some modulation of the co-evolution).

What happens in practice is that a CTA exercise, like the strategy-articulation workshops we discussed, is inserted in ongoing developments and interactions, often with support of one or more of the actors involved, for example the EU Network of Excellence Frontiers,[14] which is important to create some legitimacy for the exercise. In preparing the exercise, the organizer (CTA analyst/agent) moves about in the relevant worlds, finding out about 'entanglements', forces at play, and stakes, and using those insights to prepare for the workshop and orchestrate it. When moving about, it is the CTA analyst (as a social scientist) who inserts herself in these worlds. But in doing so, she leaves traces and thus creates small changes: the CTA analyst is already a CTA agent.

Becoming an agent in this way is not just a circumstance that requires some methodological reflection. It is actually a methodology in its own right, a methodology of insertion. Our recognizing it as a methodology emerged gradually over time. It started with the notion that the analyst moving about makes patterns in the co-evolution of technology and society visible, and thus creates some reflexivity. We learned by doing, also building on some general insights. Robinson (2010) devoted a chapter in his PhD thesis to describe his 'insertions' and their outcomes, from the perspective of a methodology in the making.

The recent interest in 'integration' or 'immersion' of social scientists and humanities scholars in the work on the lab floor can be seen as having a similar thrust, and has sometimes been developed as a methodology.[15] The important difference is that 'insertion', as we use the term here, happens at a variety of 'work floors'; it happens in a multi-layered landscape and addresses the layers explicitly. Table 3.1 indicates the layers.

The **top layer has** broad activities related to public policy, regulation, and societal debate. This includes overall institutions, arrangements, and authorities in our society.
The **middle layer** is located in collectives of actors, relevant institutions, and networks that are directly involved in nanotechnology development through coordination and agenda setting.
The **bottom layer** represents ongoing practices and projects (often shaped by enactment cycles). For nanotechnology these may occur in publicly funded research laboratories, universities, and large or small firms.

In the lab floor studies, the bottom layer is what is focussed on, but the other layers are still there, and shape what happens on the lab floor.

What does the methodology of insertion consist of? We will indicate steps, but what we mostly do is report on our learning by doing, offering some evaluations and further perspectives. The first step is 'moving about' in the world of nanotechnology. In particular, visiting locations of nanotechnology R&D, conferences and other meetings, and tracing anticipatory coordinating activities like roadmaps and European Technology Platform meetings where nanotechnology developments are being shaped. Interactions occur, and the CTA analyst & agent-to-be should be willing to enter into the substance of the developments and concerns so as to be a legitimate partner.[16] The CTA analyst must be recognized as a knowledgeable visitor, and this constitutes the second step of the methodology, the actual 'insertion' in the world of nanotechnology. Insertion is the process of becoming a temporary member of the field, a legitimate visitor. But the inserted CTA analyst should not go native, and make sure she is recognized as a visitor and not a full member.

Moving about helps to capture what is going on, and thus to target, tailor, and embed CTA exercises. CTA exercises must embed themselves, and thus fit to evolving circumstances in order to be accepted as legitimate/plausible. But there must also be some stretching of these circumstances so as to broaden enactment processes and stimulate reflexive learning. In other words, the visitor moving about is doing more than sightseeing. Fitting and stretching requires deep knowledge of dynamics and contexts. Along with the rapidly evolving developments in and around nanotechnology such knowledge can only be garnered by insertion. This is more than an anthropologist, also a visitor by definition, would do. The CTA analyst moving about in the nano-world is also formulating diagnoses about what is happening and could happen.

Insertion into the world of nanotechnology development requires the active circulation of the analyst in locations were actors are shaping the emerging paths of nanotechnology R&D. This includes research laboratories, conferences, workshops, agenda setting meetings, roadmapping events, and public debates anticipating issues related to technology developments. As a knowledgeable visitor, and based on her diagnoses of the situation, the CTA analyst can actively probe views and interactions, so as to find out about the forces at play. This will be done in preparation for a CTA exercise, but the insertion can continue over the course of a few years, so that changes over time can be traced. This is what Robinson did, within the European Network of Excellence Frontiers, and more broadly. His role evolved from 'foreigner' to 'regular': his activities became gradually accepted, visible, and in some circles, legitimate.

Important in these activities were aggregation of what was happening in the nanoworld, and analysing it, creating an overall picture, and presenting it if only in conversation with members of the nanoworld. This functions as an entry ticket ('see, I am inserted and knowledgeable') and a way of getting feedback. But there will be the danger of being positioned as part of the nanoworld, so being pressed to go native, or positioned in a service role to the nanoworld which limits the freedom of movement of the analyst. Thus, there is further requirement: play a distinct role in the nanoworld and make sure it is seen as distinct. This role of a (welcome) visitor can be highlighted by moving in and moving out of the nanoworld. The possibility to refer to own social-science publications, which could be helpful to nanoscientists and nanotechnologists (like Robinson and Propp 2008) turned out to be a good way to create legitimity. Given the vicissitudes of insertion, including working under time pressure, there will be lots of contingencies. So there will be no simple recipes.

As to overall changes, there is a clear difference between 2004 when the CTA projects started and nano-scientists looked dubiously at the intruders, and the present situation in which social scientists and other non-technical actors are welcome in the nano-world. In the particular case of Robinson, his pro-active service role was recognized, i.e. that such non-technical actors could be of some help (in indicating innovation dynamics and contributing to roadmapping, for example). The main drivers of acceptance were the pressures on the nano-world, as visible in the concerns about risk and in the call for responsible development. Listening to the knowledgeable visitor and accepting CTA exercises were ways to address these pressures.

Are outcomes in terms of CTA goals visible? Of course, it is too early to see better technology in a better society (and if so, it would not be attributable to CTA exercises). But one may see increased reflexivity in co-evolution. This relates to anticipatory coordination. In the world of nanotechnology, there is an interest in anticipation and coordination so as to choose right directions. Actual and potential stakeholders are attempting to shape emerging nanotechnology developments, in different fora and with a variety of strategies. CTA exercises are part of this move, and they create further openings. As they do this, they become recognized and accepted. There is some institutionalization of scenario/strategy workshops (Robinson 2010; Parandian 2012).

Insertion is an integral part of the CTA activities, and necessary to make them effective. It is not a means to achieve CTA goals directly, although it does contribute.[17] It is reconnoitering the lay of the land and probing the dynamics. On that basis, circumstances (like CTA workshops) can be created that stimulate actors to reflect, act, and interact in ways that might achieve the CTA agent's objectives.

A key element in achieving these objectives is making visible what was invisible to actors,[18] not by explaining (although that might occur), but in interaction with actors (that's also where the scenarios come in). As it is experimenting in real-world interactions, there is an interesting link with Lindblom's (1990) plea for inquiry rather than a search for truth as such, in relation to change. People probe the world (probe into situations, into other actor's perspectives, into problems and possible solutions) in order to change it, and this constitutes inquiry. The resulting insights can be formulated as such, somewhat independent of proposed actions. Social scientists also probe the world, whether they have a change perspective or not. Lindblom emphasizes that there is no epistemological difference between probing by citizens, by government functionaries, and by social scientists. However, as he notes, the latter may well have more honed and articulated probing skills. When one scales down the scope of Lindblom's argument from society in general to the world of nanotechnology development, it constitutes a justification of the 'insertion' approach. It is probing by the social scientist, but also stimulates probing by the actors themselves.[19]

Concluding thoughts

For new technologies, most concrete activities are at the R&D stage, rather than product development and uptake in society. Firms and research institutes are important locations, but given the open-ended promises for new technologies like nanotechnology, academic research institutions are important as well. This introduces additional dynamics, related to 'opening up the laboratory', as the title of this volume phrases it.

In a sense, scientists (even the technoscientists that abound in nanotechnology) are outsiders to society, because they live in protected places (Rip 2010c). They are insiders in their own world of science, and strongly feel like insiders, up to patrolling and protecting the boundaries of their world.[20] Bridging the gap between the inside world of science, and the outside world now occurs in various ways, pro-actively or because of outside pressures.

Social scientists and humanities scholars are outsiders to that world of science, in particular to the protected place of the lab where the work of science is done. They can visit, even become accommodated to some extent – perhaps as 'social scientist in residence'. Social scientists visiting a lab, occasionally staying there for some time, shift out of their own world. Anthropologists and ethnographers (of science) have been doing that all along, but with another purpose, to gather data rather than changing the world they study. Their presence would increase reflexivity of the actors, however, whether they wanted that or not. Our methodology of insertion is explicit about this.[21]

CTA has a larger scope, and addresses embedding in society, if only through anticipation. The dynamics will be more complex: there are now different overlapping worlds, different perspectives, and actors at the collective level (ranging from branch organizations to government agencies), with some collective responsibility. And there are larger and long-term developments, in particular the traditional division of labour between promotion and control, which is now questioned, as in the discourse of responsible research and innovation.

Concretely, in the world of nanotechnology, CTA exercises are welcomed (and funded) by the technology developers and technology promoters, who see them as necessary to anticipate societal embedding, and meeting possible reactions from various societal actors. Co-evolution of technology and society goes on anyway, but anticipations are becoming more important, so that the co-evolution will be more reflexive – even if enactors will work from their concentric perspective.

If co-evolution becomes reflexive, and actors absorb CTA activities in their practices, will CTA agents become superfluous? Not yet, and probably never. One reason is that CTA agents can circulate across locations and observe and analyse what happens at the collective level, which will be more difficult for regular actors. Another reason is that these visiting 'knowledgeable' strangers irritate existing ways of working and thus create openings for learning and further evolution of how we handle new technologies in our society.

Notes

1 This then led technology developers to see TA as 'technology harassment'.
2 We note that there is another tradition of TA, in firms and research institutes, where technological options are assessed as to eventual performance and production possibilities and costs. This can be called 'technical' TA, to distinguish it from the 'public' TA that we discussed here (Rip 2001a). When broader considerations would be taken into account, 'technical' TA would become 'sociotechnical' TA, and the tools of CTA (see below) could be used by the firms and research institutes, or by consultancies that are commissioned to do 'sociotechnical' TA.
3 Note that 'reflexivity' here refers to institutions and approaches in society and sectors in society, not to individuals becoming more reflective – even while that is part of overall reflexivity.
4 A well-known precedent is the temporary moratorium on recombinant DNA research, after the 1974 Asilomar meeting. The present call for a moratorium on nano-particle development comes from critical outsiders, not from nanoscientists. A mixed case (early 2012) is the voluntary stop (for 60 days) of bird flu virus research, after the US National Science Advisory Board on Biosecurity had required a virology research group in Erasmus University Rotterdam to take out details in their pending publication in *Science*, because of the risk of misuse.
5 Perspectives, expectations, preferences, and positions of various actors/stakeholders will be articulated, i.e. become more explicit, further specified and linked to arguments, findings, and values, in interaction and this may lead to scrutiny and assessment.
6 This is particularly important when the focus is on embedding of technology in society (including further sociotechnical development). This is how Hoogma and Schot evaluated social experiments with electric vehicles (Hoogma 2000; Hoogma et al. 2002; see also Schot and Rip 1997).
7 Ethical, Legal, and Societal Aspects, the 'Aspects' are sometimes referred to as Issues (then the acronym becomes ELSI).
8 The term 'enactor' can be used for all cases where a project is pursued, and identification occurs so that the world is seen in terms of whether it helps or hinders the project. An actor can be enactor in one case, and comparative selector in another case.

An interesting example is the NGO Greenpeace, almost by definition an outsider/comparative selector. But Greenpeace Germany, at one moment, pushed for an environmentally-friendly fridge, and collaborated with scientists and a firm to realize it (Van de Poel 1998: 84–97). So it became an enactor, for the time being.

9 This is often a novel possibility for participants. Moving beyond their own interests and perspectives comes easier to some than others, but it is recognized as a possibility in post-workshop interviews with participants (Parandian 2012). The set-up of a CTA workshop has to facilitate and stimulate this, by making sure various actor perspectives are visible, and possible developments in the real world are considered, for example with the help of sociotechnical scenarios.

10 Marris et al. (2008) have shown this for an Interactive TA exercise about field tests of genetically modified vines in France. Their point is reinforced by what happened subsequently: productive co-construction of the design of the field tests between local stakeholders and researchers, and five years later, August 2010, the destruction of the test fields by critics of GMO. In LMC et al. (2010), the story is told from the perspective of the actors involved in the co-construction.

11 Scenarios add substance to the interactions, which is necessary because they are not just about participation and empowerment (which are sometimes taken as goals for CTA, cf. earlier comments on democracy). To serve the change aim of CTA, they must be seen as relevant as well as challenging to the participants. Quite some effort has to be put into the creation of robust socio-technical scenarios. Thus, they become a product in their own right, which can be put to further use, also by participants.

12 Versions of the hype-cycle were presented by Gartner Group since at least 1999, see Fenn, (1999).

13 The emphasis on choices in ongoing developments is also important to counter the opposite position, that there is no way to predict future impacts of a technology, so better give up on technology assessment and other attempts at anticipation and feedback. This 'hard truth' was pushed by Nathan Rosenberg in an OECD workshop on Social Sciences and Innovation (Tokyo, 2000), but it overlooks how present dynamics shape opportunities and constraints for future developments, and are thus a basis for anticipation and feedback (Rip 2001b). The further point is that anticipations need not be correct to be useful in guiding action – think of self-negating prophecies.

14 This network of nanotechnology research institutions focussed on the development of nanotechnology instrumentation and approaches for the life sciences (see Robinson 2010).

15 In particular in the Socio-Technical Integration Research (STIR) project, funded by the US National Science Foundation and led by Erik Fisher (Arizona State University). See Schuurbiers and Fisher (2009).

16 So this is more than participant observation, or anthropologists alternating between insider and outsider positions.

17 Social scientists moving about in the world of a scientific specialty or domain will set the members of that world thinking about what is happening, and about patterns that enable or constrain. This is relevant for the overall CTA goal of increasing reflexivity of co-evolution of technology and society. Moving about in the nano-worlds may have such an effect, but it was not an explicit aim that structured the moving about.

18 A sort of sociological enlightenment in the small, cf. Rip and Groen (2001).

19 Phrased in this way, there is overlap with participatory research approaches (cf. Bergold and Thomas 2012). There, the social scientists have the higher status, while in our case, nano-scientists and policy makers tend to relegate the social scientists to a service role. Thus, building trust will have a different complexion.

20 There is a functional argument: scientists should live in protected spaces, at least to some extent, in order to be productive (Rip 2010c).

21 There are normative issues involved, which can refer to the background goals of CTA, but also have an experimental component, finding out about the issues by doing and learning (cf. also Laurent and Van Oudheusden, forthcoming).

References

Bergold, J. and Thomas, S. (2012), Participatory research methods: A methodological approach in motion. *Forum: Qualitative Social Research* 13(1), Article 30.

Callon, M., Lascoumes, P. and Barthe, Y. (2001), *Agir dans un monde incertain. Essai sur la démocratie technique*. Paris: éd. Seuil.

Daey Ouwens, C., Hoogstraten, P. van, Jelsma, J., Prakke, F. and Rip, A. (1987), *Constructief Technologisch Aspectenonderzoek. Een Verkenning*. Den Haag: Staatsuitgeverij (NOTA Voorstudie 4).

Delvenne, P. (2011), *Science, technologie et innovation sur le chemin de la réflexivité. Enjeux et dynamiques du Technology Assessment parlementaire*. Louvain-La-Neuve: Harmattan-Academia.

European Commission (2008), Commission recommendation on a code of conduct for responsible nanosciences and nanotechnologies research. *C(2008)424 Final* (7 February).

Fenn, J. (1999), *When to Leap on the Hype Cycle: Research Note*. Stamford, CT: Gartner Group. www.cata.ca/files/PDF/Resource_Centres/hightech/reports/indepstudies/When toleaponthehypecycle.pdf

Garud, R. and Ahlstrom, D. (1997), Technology assessment: A socio-cognitive perspective. *Journal of Engineering and Technology Management* 14, pp. 25–48.

Genus, A. (2006), Rethinking constructive technology assessment as democratic, reflective, discourse. *Technology Forecasting & Social Change* 73, pp. 13–26.

Grin, J. and van de Graaf, H. (1996), Technology assessment as learning. *Science, Technology, & Human Values* 21, pp. 72–99.

Guston, D.H. and Sarewitz, D. (2002), Real-time technology assessment. *Technology and Society* 24, pp. 93–109.

Harremoës, P. (ed.) (2001), *Late Lessons From Early Warnings: The Precautionary Principle 1896–2000*. Copenhagen: European Environment Agency.

Hoogma, R. (2000), *Exploiting Technological Niches: Strategies for Experimental Introduction of Electric Vehicles*. PhD dissertation, Enschede: Twente University Press.

Hoogma, R., Kemp, R., Schot, J. and Truffer, B. (2002), *Experimenting for Sustainable Transport: The Approach of Strategic Niche Management*. London: Spon Press.

Laurent, B. and van Oudheusden, M. (forthcoming), Experimental normativity: Shifting and deepening engagement in public participation in science and technology. *Science as Culture*.

Lindblom, C.E. (1990), *Inquiry and Change: The Troubled Attempt to Understand and Shape Society*. New Haven, CT: Yale University Press.

LMC et al. (Local Monitoring Committee, Olivier Lemaire, Anne Moneyron, Jean E. Masson) (2010), "Interactive technology assessment" and beyond: The field trial of genetically modified grapevines at INRA-Colmar. *PLoS Biology* 8(11) (November) pp. 1–7, separately paginated.

Marris, C., Rip, A. and Joly, P.-B. (2008), Interactive technology assessment in the real world: Dual dynamics in an iTA exercise on genetically modified vines. *Science, Technology & Human Values* 33(1), pp. 77–100.

Ministerie van Onderwijs en Wetenschappen. *Integratie van Wetenschap en Technologie in de Samenleving. Beleidsnota*. 's-Gravenhage: Tweede Kamer 1983–1984, 18 421(1–2). (Policy Memorandum: Integration of Science and Technology in Society).

Nordmann, A. and Rip, A. (2009), Mind the gap revisited. *Nature Nanotechnology* 4(May) pp. 273–274.

Parandian, A. (2012), *Constructive TA of Newly Emerging Technologies: Stimulating Learning by Anticipation Through Bridging Events*. PhD thesis, Technical University Delft, defended 12 March.

Parandian, A., Rip, A. and te Kulve, H. (2012), Dual dynamics of promises and waiting games around emerging nanotechnologies. *Technology Analysis & Strategic Management* 24(6), pp. 565–582.

Rip, A. (1992), Between innovation and evaluation: Sociology of technology applied to technology policy and technology assessment. *RISESST* 2, pp. 39–68.

Rip, A. (2001a), Technology assessment. In N.J. Smelser and P.B. Baltes (eds.), *International Encyclopedia of the Social & Behavioral Sciences*, Vol. 23. Oxford: Pergamon Press, pp. 15512–15515.

Rip, A. (2001b), *Assessing the Impacts of Innovation: New Developments in Technology Assessment*. In OECD Proceedings, Social Sciences and Innovation, Paris, pp. 197–213.

Rip, A. (2002), *Co-Evolution of Science, Technology and Society*. Expert Review for the Bundesministerium Bildung and Forschung's Förderinitiative 'Politik, Wissenschaft und Gesellschaft' (Science Policy Studies), managed by the Berlin-Brandenburgische Akademie der Wissenschaften. Enschede: University of Twente (7 June).

Rip, A. (2006), Folk Theories of Nanotechnologists. *Science as Culture* 15(4) (December 2006), pp. 349–365.

Rip, A. (2009), Futures of ELSA. *EMBO Reports* 10(7), pp. 666–670.

Rip, A. (2010a), De facto governance of nanotechnologies. In M. Goodwin, B.-J. Koops and R. Leenes (eds.), *Dimensions of Technology Regulation*. Nijmegen: Wolf Legal Publishers, pp. 285–308.

Rip, A. (2010b), Processes of entanglement. In M. Akrich, Y. Barthe and F. Muniesa et Philippe Mustar (réd.), *Débordements. Mélanges offerts à Michel Callon*. Paris: Transvalor – Presses des Mines, pp. 381–392.

Rip, A. (2010c), Protected spaces of science: Their emergence and further evolution in a changing world. In M. Carrier and A. Nordmann (eds.), *Science in the Context of Application: Methodological Change, Conceptual Transformation, Cultural Reorientation*. Dordrecht: Springer, pp. 197–220.

Rip, A. and Groen, A. (2001), Many visible hands. In Rod Coombs, Ken Green, Vivien Walsh and Albert Richards (eds.), *Technology and the Market: Demands, Users and Innovation*. Cheltenham: Edward Elgar, pp. 12–37.

Rip, A., Misa, T.J. and Schot, J. (eds.) (1995), *Managing Technology in Society: The Approach of Constructive Technology Assessment*. London and New York: Pinter Publishers.

Rip, A. and Shelley-Egan, C. (2010), Positions and responsibilities in the "real" world of nanotechnology. In René von Schomberg and Sarah Davies (eds.), *Understanding Public Debate on Nanotechnologies: Options for Framing Public Policy*. Brussels: Commission of the European Communities, pp. 31–38 (January).

Rip, A. and te Kulve, H. (2008), Constructive technology assessment and sociotechnical scenarios. In E. Fisher, C. Selin and J.M. Wetmore (eds.), *The Yearbook of Nanotechnology in Society, Volume 1: Presenting Futures*. Berlin etc: Springer, pp. 49–70.

Robinson, D.K.R. (2009), Co-evolutionary scenarios: An application to prospecting futures of the responsible development of nanotechnology. *Technological Forecasting & Social Change* 76, pp. 1222–1239.

Robinson, D.K.R. (2010), *Constructive Technology Assessment of Emerging Nanotechnologies: Experiments in Interactions*. PhD thesis, Enschede: University of Twente, defended 25 November.

Robinson, D.K.R. and Propp, T. (2008), Multi-path mapping as a tool for reflexive alignment in emerging S&T. *Technological Forecasting and Social Change* 75, pp. 517–538.

Schot, J. and Rip, A. (1997), The past and future of constructive technology assessment. *Technological Forecasting and Social Change* 54, pp. 251–268.

Schuurbiers, D. and Fisher, E. (2009), Lab-scale intervention. *EMBO Reports* 10(5), pp. 424–427.

Swierstra, T. and Rip, A. (2007), Nano-ethics as NEST-ethics: Patterns of moral argumentation about new and emerging science and technology. *NanoEthics* 1 pp. 3–20.
Te Kulve, H. (2011), *Anticipatory Interventions in the Co-evolution of Nanotechnology and Society*. PhD thesis, University of Twente, defended 21 April.
Van de Poel, I. (1998), *Changing Technologies: A Comparative Study of Eight Processes of Transformation of Technological Regimes*. Enschede: Twente University Press.
Von Schomberg, R. (2007), *From the Ethics of Technology Towards and Ethics of Knowledge Policy*. Working Document of the Service of the European Commission. http://ec.europa.eu/research/science-society/pdf/ethicsofknowledgepolicy_en.pdf

Index

Note: Page numbers in *italics* indicate figures, those in **bold** indicate tables, and those with n indicate notes.

Milton Keynes UK
Ingram Content Group UK Ltd.
UKHW040051071024
449327UK00019B/491

9 780367 786205